Sharpening Your
SAS
Skills

T0262789

Sharpening Your

SAS

Skills

Sunil Gupta
Curt Edmonds

CH Chapman & Hall/CRC
Taylor & Francis Group

Boca Raton London New York Singapore

Published in 2005 by
Chapman & Hall/CRC
Taylor & Francis Group
6000 Broken Sound Parkway NW, Suite 300
Boca Raton, FL 33487-2742

Library of Congress Cataloging-in-Publication Data

Gupta, Sunil, 1963-
    Sharpening your SAS skills / by Sunil Gupta and Curt Edmonds.
        p. cm.
    Includes bibliographical references and index.
    ISBN 1-58488-501-7 (alk. paper)
    1. SAS (Computer file) 2. Mathematical statistics--Data processing. I. Edmonds, Curt. II. Title.

QA276.4. G88 2005
519.5'0285--dc22   2004063431

Taylor & Francis Group
is the Academic Division of T&F Informa plc.

Visit the Taylor & Francis Web site at
http://www.taylorandfrancis.com

and the CRC Press Web site at
http://www.crcpress.com

Preface

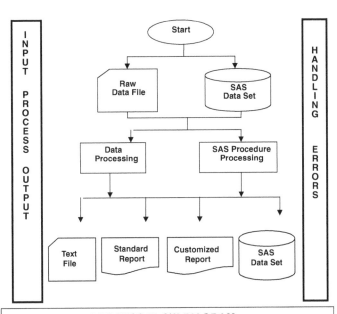

PROCESS FLOW DIAGRAM

PHASE	CHAPTER, DESCRIPTION
INPUT:	1. Accessing Data
	2. Creating Data Structures
PROCESS:	3. Managing and Summarizing Data
OUTPUT:	4. Generating Reports
HANDLING ERROR:	5. Diagnosing and Correcting Errors
V8.2/9.1:	6. Integrity Constraints, Generation DS, Audit Trials

Sharpening Your SAS Skills is a concise guide designed to help you read, understand, and write better SAS programs. The book's logical organization and detailed comparison of key facts make it unique. Examples presented in this book help you to diagnose the most common SAS programming tasks in data access, data management, data analysis, and data presentation. The book prioritizes and reviews information on topics such as the DATA step versus the SQL procedure, the rules for common variables, and the sequence order for DROP and KEEP statements. For a better understanding of the SAS programming language, the book focuses on the similarities and differences between SAS syntax and programming approaches. Each example illustrates the concept and shows the results expected from the input data. ODS examples are included to show how easy it is to create HTML, PDF, and RTF files. In addition, SAS Version 9.1 examples are included to show how to improve the quality of data entered in data sets. Finally, there are over 150 technical questions to test your SAS knowledge.

Because it is not possible to list and describe all SAS statements or options in this book, references are made to SAS papers, manuals, books, and web sites for more information. The intent of this book is not to be a complete SAS reference book, but to establish a strong understanding of the SAS programming language by reviewing and testing your knowledge of the SAS programming essentials.

The process flow diagrams in this book illustrate the four main components of SAS programming: input, process, output, and handling errors. Each chapter goes into detail about how SAS is used to accomplish that phase, from reading the raw data to creating SAS data sets to data processing with the DATA step and SAS procedures to creating text files, reports, and output data sets. Since errors can occur throughout the process, a separate chapter is dedicated to handling errors. Chapter 6 reviews enhancements in SAS Version 8.2 and Version 9.1 that affect all four components. Where possible, a general rules section is presented first within each section to establish common syntax rules.

The symbols and terms below are used in the process flow diagram and are defined as follows:

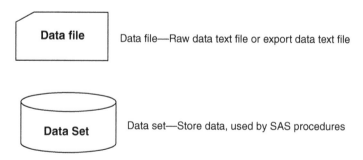

Data file   Data file—Raw data text file or export data text file

Data Set   Data set—Store data, used by SAS procedures

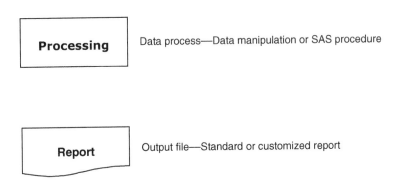

Processing   Data process—Data manipulation or SAS procedure

Report   Output file—Standard or customized report

Output file—standard or customized report

Chapter 1 introduces methods for accessing data from external files. Variations of the INPUT statement are used to create SAS data sets from most any data layout. Once data sets are created, they can be merged together to combine the data. One of the comparisons made in this chapter is the difference between the LIST, COLUMN, and FORMATTED INPUT statements.

Chapter 2 provides examples on how to subset records and select variables. Other examples show how date variables are created and used in calculations. Data from data sets can be exported to an external file for reporting or further processing. One of the comparisons made in this chapter is the difference between the WHERE and IF statements when subsetting data sets. In addition, a detailed step-by-step analysis of the compile and execution phases is reviewed.

Chapter 3 provides numerous examples for managing data using the DATA step. Everything from creating variable attributes, conditionally executing SAS code, using SAS functions, and accumulating totals is shown. Processing data using DO loops and SAS arrays improves program efficiency. One of the comparisons made in this chapter is the difference in DO loops as far as knowing the rules for top or bottom evaluation of the expression.

Chapter 4 shows how reports and output can be produced from SAS procedures. Basic syntax of the PRINT, MEANS, FREQ, TABULATE, and REPORT procedures are introduced. Summary and detailed level reports using ODS to create HTML, PDF, and RTF files will also be shown. Data can also be exported to Excel.

Chapter 5 shows how to handle ERRORs, WARNINGs, and NOTEs in SAS logs. Most every SAS user experiences these messages that need to be identified, diagnosed, and resolved. A variety of ERROR and WARNING messages is generated from the examples to show the scope of the type of messages generated. This chapter helps you better to understand the reason for the message so that ERRORs and WARNINGs can be prevented.

Chapter 6 covers the enhancements in SAS Version 8.2 and Version 9.1. SAS users can take advantage of the new 9.1 features for preventing data entry errors, tracking updates with audit trails, and backing up data with generation data sets. In addition, new SAS functions facilitate character and numeric data manipulation.

The question section at the end of each chapter include technically challenging questions to reinforce your SAS skills. Answers are located at the end of the book. The appendix section includes a brief description of key terms. In addition, these SAS keywords and terms are upper case in the content for ease of reference. The reference section includes a thorough list of good SAS papers, books, and manuals for more detailed information on a specific topic.

All SAS output files, listing outputs, and log files are generated with SAS Release 8.2 and 9.1 on the Windows XP operating system using SAS/BASE. Each SAS program and output is annotated with numbers for ease of reference in the book content. For the IMPORT and EXPORT procedures to access PC files directly, as shown in Chapter 1 and Chapter 2, SAS/ACCESS to PC is also required. All examples are expected to work in batch mode and from the program editor of the SAS windowing environment.

For those new to SAS, all of the examples in this book can be practiced using SAS's Learning Edition software. Instead of using the default interface to select tasks, you can directly enter SAS code with the insert code option and view the results.

About the Authors

Sunil Gupta is a principal consultant at Gupta Programming (www. GuptaProgramming.com). He has been using SAS® software for over 11 years and is a SAS Base Certified Professional. Gupta is also the author of *Quick Results with the Output Delivery System*, developer of over 5 SAS programming classes, developer of Clinical Trial Reporting Templates for quick generation of tables, lists and graphs and was a SAS Institute Quality Partner™ for over 5 years.

Curt Edmonds is a managing partner of SimulStat Incorporated (www. simulstat.com), specializing in SAS programming for the pharmaceutical industry. He has been using SAS® software for over 10 years and has attended numerous SAS® training classes. Edmonds has been a SAS® consultant with various pharmaceutical companies in all phases of drug development and has participated in several successful FDA submissions.

Acknowledgments

As Curt and I come to end of writing our book *Sharpening Your SAS** *Skills*, we would like to thank our family, friends, and associates for their excitement, dedication, and support. We realize that this accomplishment would not have been possible without the time and interest of everyone involved.

We would like to give special thanks to our family members: Bindiya and Michelle for their encouragement and for letting us work those long, long hours; and Aarti and Anupama for their energy and excitement. We would also like to thank our peer review team members for their attention to detail and their great suggestions: Adam Sharp, Mary Katz, John Hunter, Brian Sheehan, Deepak Asrani, Karthikeyan Chidambaram, and Charles Patridge. Thanks to the staff at Taylor & Francis for their assistance and support: Rob Calver, Kristy Stroud, Tao Woolfe, Jonathan Pennell, Katy Smith, Clare Brannigan, and Theresa Del Forn.

We hope that all SAS users will benefit from the organization and content of this book as they sharpen their SAS skills.

*SAS is a trademark of SAS Institute Inc. in Cary, North Carolina. Contact them at www.sas.com.

Contents

Chapter 1

Accessing Data

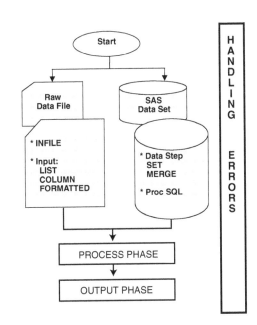

1.1   Introduction

One of the important features of SAS is the ability to create SAS data sets from raw data. Raw data is data stored in an external file or listed after a CARDS or DATALINES statement within the SAS program. This chapter focuses on the different ways to read raw data: as LIST, COLUMN, FORMATTED, or NAMED INPUT. The main syntax in SAS to read raw data is the INPUT statement. This chapter describes its use for each of the methods along with the different components: column and line pointer controls and trailing @ controls. The last section of this chapter focuses on using the DATA step and the SQL procedure to combine SAS data sets. In general, when creating data sets, it is useful to use the PRINT and CONTENTS procedures to verify the data read.

1.2   LIST, COLUMN, FORMATTED and NAMED INPUT to Read Raw Data

1.2.1   General Rules

When using the LIST, COLUMN, FORMATTED, and NAMED INPUT statement to read raw data, there are several general rules that apply to all INPUT statements. The INPUT statement is used to create the specified variables. When creating variables, **numeric data (data containing only numbers, decimal points, or minus signs) will become right justified and character data (letters, numbers, or special characters) will become left justified** in the new data set. A dollar sign ($) is used after the variable name to identify a character variable type. In addition, **character data is case-sensitive.** For example, the character data "Tim" is not equal to "tim." The maximum number of variables that can be created in a data set is 32,767.

1.2.1.1   LIST INPUT Statement

Syntax:
INPUT <variable> $;

The syntax of the LIST INPUT statement is made up of the INPUT keyword followed by the list of variables. When using the LIST INPUT statement, several things should be kept in mind regarding the data. First, the LIST INPUT statement will read all the variables from the left side to the right side in the order that they are specified in the statement. The raw data must be in the same order as the variables and must be separated by a blank or a delimiter. Although the default delimiter is the comma, it can be changed to any character symbol. Second, with the LIST INPUT statement, **variables cannot be skipped or reread.** Thus, you cannot read only the variables that you need and you cannot read the variables more than once.

Even though the LIST INPUT statement is the simplest of the input types, it is restrictive in the type and format of data that can be read. One of the biggest limitations is that the data must be in **standard character or numeric** format. Formats are used by SAS to display data values. Informats are similar to formats except that they are used to read data values. The standard format is 8. for numeric variables and $8. for character variables. Since an informat is required to read dates, the basic list input statement will not read date values. Thus, all variable lengths will be 8 bytes unless the LENGTH statement is specified. See Section 2.3.4 for more information on dates and Chapter 3, Section 3.2 for more information about variable attributes.

In regard to the data, if there is a missing value for one of the variables, then a placeholder must be used. **For both numeric and character variables, a period represents missing data.** In the case of missing numeric variables, SAS will treat this as the smallest numeric value when sorted. One other restriction for character variables is that the LIST INPUT statement, by default, will **not allow embedded blanks** in the data; e.g., "my house" is not allowed as a character variable value.

To get around some of the LIST INPUT limitations, format modifiers can be applied. For example, the INPUT DOB : MMDDYY8. statement can be used to read the date variable DOB in the MMDDYY8 informat. The additional benefit of the colon (:) format modifer is that the raw data does not have to be lined up in specific columns. Table 1.1 shows the three types of format modifiers that are available with the LIST INPUT statement.

Table 1.1 Format Modifiers

Format Modifier	Example	Description
Ampersand (&)	INPUT x y & $7.; data: 1 one two	Used to read character values that might contain single embedded blanks.
Colon (:)	INPUT x y : $7.; data: 1 one	Reads nonstandard data values and character values that are longer than eight characters.
	INPUT DOB: MMDDYY8.; data: 10/23/78	Scans for the next data value instead of expecting only one space between data values. Ends reading when either a space is read, the length of the variable (if character), or the end of the raw data line is reached.
Tilde (~)	INFILE datfile DSD; INPUT bktitle ~ $11.; data: "Peter Pan"	Used with the DSD INFILE statement option to cause quote marks to be treated as valid data.

Example 1.1   Read data using the LIST INPUT statement.

```
data class;
            ❶                ❷
input name $ absences quiz class weight;


/* Column Reference Number
123456789012345678901234567890 */
cards;         ❸
tim 1 6.0 3 .
sally 2 10.0 4 +110
john -2 8.0 3 120

;
run;


proc print data=class;
run;


proc contents data=class;
run;
```

Output

```
The SAS System
Obs   name     absences   quiz   class   weight
       ❹                    ❺
  1    tim         1         6      3        .
  2    sally       2        10      4       110
  3    john       -2         8      3       120
```

Contents

```
-----Alphabetic List of Variables and Attributes-----
        #    Variable    Type    Len         Pos

        2    absences    Num      8      ❻     0
        4    class       Num      8           16
        1    name        Char     8      ❼    32
        3    quiz        Num      8            8
        5    weight      Num      8           24
```

In Example 1.1, the *name, absences, quiz, class,* and *weight* variables are listed after the INPUT statement and appear in the **same order as the raw data.** The $ after the variable *name* ❶ indicates that it is a character variable, while all of the other variables are numeric ❷. The CARDS or DATALINES statement is used to signal SAS that the raw data being read is listed on the next line. When looking at the raw data, you will notice the values are separated by a blank and that the weight variable is missing for the first observation ❸. A semicolon is used on the line following the last line of data to tell SAS that there is no more data to be read. When using the LIST INPUT statement, data values must be separated by blanks or a delimiter. In the data, it is acceptable to use positive or negative signs for numeric data.

Applying the PRINT procedure on the data set will show if the data has been read correctly in the data set. In the output window the character variable *name* is left justified with **no embedded blanks** ❹. The numeric variables are all right justified ❺ with missing values displayed as a period.

Applying the CONTENTS procedure on the data sets will show all of the variables in the data set along with their attributes. When using the LIST INPUT statement, you will notice that **by default, the length of all variables is 8 bytes.** The informat and format of all the numeric variables is 8. ❻ and the informat and format of all character variables is $8. ❼.

1.2.1.2   COLUMN INPUT Statement

Syntax:
INPUT <variable> $ <column start – column end>;

The COLUMN INPUT statement is used to read data when the data values are always located in the same position. The syntax of the COLUMN INPUT statement is similar to that of the LIST INPUT statement with one exception, the addition of the column identifier. The column identifier specifies the starting and ending columns of each variable.

The COLUMN format has several advantages over the LIST format. One of the biggest advantages is that **no delimiter is needed to separate the data** in the COLUMN format, thus saving space and decreasing the size of your input file. There are also advantages in the type of data that can be read with the COLUMN INPUT statement. Unlike the LIST INPUT statement, the data is not limited by the 8 byte default but can be of any length from 1 to 32,767 characters. **The length of character variables is determined from the total number of columns between the start and end columns.** In addition, **character data can contain embedded blanks;** e.g., "my house" is allowed in the data. **As with the LIST INPUT statement, the default length for numeric variables is 8 bytes unless the LENGTH statement is specified.**

Also, the way that COLUMN INPUT statement reads data is an important improvement over the way the LIST INPUT statement reads data. With the COLUMN INPUT statement, **you do not have to create a placeholder in your raw data for missing numeric or character data or read all of your data.** Since SAS reads variables as specified by the starting and ending columns, **variables may be skipped over** by not specifying them in the INPUT statement. You can also **reread data** in the COLUMN INPUT statement by specifying the same starting and ending columns. This process saves the data to multiple variables. By changing the ending or starting columns, you can cause portions of the data to be read into a variable or omitted from the input record.

Example 1.2   Read data using the COLUMN INPUT statement.
```
data class2;        ❶        ❸                            ❹
 input name $ 1-7 absences 9-10 quiz 11-14 weight 17-20
 class 15;
/* Column Reference Number
12345678901234567890123456789 */
cards;
tim lou  1 6.03 .  ❺
sally    210.04 +110
john     -2 8.03 120
;
proc print data=class2;
run;

proc contents data=class2;
run;
```

Output

```
Obs   name        absences   quiz    weight   class

 1    tim lou        1         6        .        3
 2    sally          2        10       110       4
 3    john          -2         8       120       3
```

Contents
```
-----Alphabetic List of Variables and Attributes-----
          #      Variable    Type    Len       Pos

          2      absences    Num      8          0
          4      class       Num      8         16
          1      name        Char     7   ❷     32
          3      quiz        Num      8          8
          5      weight      Num      8         24
```

In Example 1.2, the *name, absences, quiz, class,* and *weight* variables are listed after the INPUT statement with the ranges of the columns for each variable. The *name* variable will store the data from columns 1 to 7 ❶. Note that the length of the *name* variable is now 7 instead of 8. ❷ All other numeric variables are still 8 bytes. Because this is a range, **embedded characters are allowed (e.g., "tim lou").** In addition, ranges also define the informat which contains the instructions on how the data will be read. The *absences* variable will have the 2. informat applied instead of the default 8. informat, one for the position at column 9 and another for the position at column 10 ❸. Notice that columns 17 to 20 were read before column 15 ❹ because the *weight* variable was placed before the *class* variable. The *weight* variable will be set to missing ❺ since the value for the variable was a dot. Note that the dot is not required since using the COLUMN INPUT statement.

1.2.1.3 FORMATTED INPUT Statement

Syntax:
INPUT @ <column number> <variable> <informat>;

The FORMATTED INPUT statement can be used to **read both standard and nonstandard data. An informat is required for the data to be read correctly**. Standard data is data that contains only numbers, decimal points, and a positive or negative sign. A number that has a comma or a dollar sign would be considered nonstandard data and needs to be read with a FORMATTED INPUT statement. **A special type of nonstandard data is dates.** SAS will read a date and convert it to the number of days from January 1, 1960 to the given date. The date will then be stored as a standard numeric variable. Another type of special nonstandard data is times. SAS will read a time and convert it to the number of seconds from midnight. The length of data and time variables is always 8 bytes.

The syntax of the FORMATTED INPUT statement is a little different from what you have seen so far. As with the LIST and COLUMN INPUT statements, the syntax starts with the INPUT keyword. Following the keyword, you can use the syntax of the LIST (e.g., "<variable> $") or COLUMN (e.g., "<variable> $ 1-7") statement or as in the syntax above you can use the column absolute pointer control @. The relative pointer control + can also be used. This provides the greatest flexibility when reading data. The column pointer control is probably the best way, since SAS will read the column as specified by the starting point up to the width of the informat. Following the variable name is the informat that will be used to read the data. In general, the informat will be followed by a period. The only exceptions to this rule are informats that use a decimal point; e.g., comma12.2. Note that it is very important to specify a period in the informat to prevent SAS from interpreting the informat as a variable name and trying to create this variable.

Example 1.3   Read data using the FORMATTED INPUT statement.

```
data class3;
              ❶      ❷              ❸              ❹
   input name $7. @10 absences 2. @15 quiz 4.1 @21 class
                      ❺
   1. @23 dob mmddyy8.;
/* Column Reference Number
12345678901234567890123456789 */
cards;
tim lou  1     6.0   3   10/23/78
sally    2    10.1   2   01/02/78
john     2     8.0   3   06/22/78
;
run;

proc print data=class3;
run;
proc contents data=class3;
run;
```

Output

```
The SAS System
                                                              ❻
Obs        name           absences     quiz     class    dob

 1         tim lou            1          6.0       3      6870
 2         sally              2         10.1       2      6576
 3         john               2          8.0       3      6747
```

Contents

```
-----Alphabetic List of Variables and Attributes-----

        #        Variable     Type     Len      Pos

        2        absences     Num       8        0
        4        class        Num       8        16
        5        dob          Num       8        24
        1        name         Char      7        32
        3        quiz         Num       8        8
```

In Example 1.3, the *name*, *absences*, *quiz*, *class*, and *dob* variables are created. The *name* variable ❶ is created with LIST INPUT, as the variable is specified after the INPUT statement along with the *$* to designate a character variable and the length of 7. The *absences* variable ❷ is created using the column pointer control @. The pointer control will cause the pointer to move to column 10 to start reading the data. The pointer will then stop reading at column 11 since the informat is 2 bytes long ❸. The *quiz* variable ❹ is created with an informat that has a decimal. The raw data already has a decimal point so the informat of 4.1 is used. The last variable *dob* is an **example of nonstandard data.** Because date values contain the slash (/), the MMDDYY8. informat is used ❺. The pointer control will start reading from column 23 and read the data until the last column indicated by the format is reached. **Dates are represented as the number of days since January 1, 1960,** so the value of the *dob* variable will be a numeric value ❻. **Note that the length of all numeric variables is still 8 bytes. Note also that without a FORMAT statement to display date variables, it becomes difficult to understand dates.** See Chapter 3, Section 3.2 for more information on the FORMAT statement.

1.2.1.4 NAMED INPUT Statement

Syntax:
INPUT @ 'text' <variable> format;
     or
INPUT text=$;

The NAMED INPUT statement is used to read raw data that contains a specified text. This can be done in two different ways. The first method is similar to that of the FORMATTED INPUT statement, but instead of specifying the column number, the column pointer control and a text string is used instead. The first example below uses the specified text format.

Example 1.4 Read data using the NAMED INPUT statement with the @'text' option.

```
data logstats;
  Input @'used' duration 6.2;
cards;
used 0.05
❶ used 2.72
run;


proc print data=logstats;
run;
```

Output

```
Obs   duration
 1      0.05
 2      2.72
```

In Example 1.4, the *duration* variable is created by searching the data for the keyword "used." When the column pointer control is used with the "text" syntax, SAS searches for the "text" and positions the pointer to start reading the raw data following the "text." The data will then be read into the variable using the specified informat. Note that the "text" in the raw file does not need to be in the same column ❶.

The second way to use the NAMED INPUT statement is to use the "text = $" syntax. This type of statement is used to read raw data that contains specified text followed by the equal sign.

Example 1.5   Read data using the NAMED INPUT statement with the text = $ option.

```
data car;
  input car = $;
cards;
car = toyota
car = van          ❶
run;


proc print data=car;
run;
```

Output

```
Obs car
1 toyota
2 van
```

In Example 1.5, the *car* variable is created as a character variable and contains the data that appears in the raw data after the equal sign ❶. Since the length is not specified, the default length of 8 bytes will be applied.

1.3   Using Various Components of an INPUT Statement

There are additional components that can be used with the INPUT statement. In addition to the column pointer controls, the INPUT statement also has these components: line pointer controls, trailing @ controls, and double trailing @@ controls. This section introduces you to the different types of controls and how they can be used to read raw data.

1.3.1   Column Pointer Control @

Syntax:
INPUT @<column number> <variable> informat;

You have already seen the use of column pointer controls in the FORMAT-TED INPUT statement. The column pointer control works best with the

FORMATTED INPUT statement and an informat. The column pointer control is used **to place the pointer at a specified column.** SAS will read as many characters as specified by the informat. By default, the pointer will start reading the raw data at the first column, so "@1" is not needed.

Example 1.6   Read data using the column pointer control.

```
data class3;
                 ❶
input name $7. @10 absences 2. @15 quiz 4.1 @21 class
1. @23 dob mmddyy8.;
/* Column Reference Number
12345678901234567890123456789 */
cards;
tim lou  1     6.0    3 10/23/78
sally    2    10.1    2 01/02/78
john     2     8.0    3 06/22/78
;
run;

proc print data=class3;
run;
```

Output

```
The SAS System
```

Obs	name	absences	quiz	class	dob
1	tim lou	1	6.0	3	6870
2	sally	2	10.1	2	6576
3	john	2	8.0	3	6747

In Example 1.6, the column pointer is not needed to read the *name* variable since it is located in the first column and SAS will start at column 1 by default. After reading the *name* variable, **the pointer will be placed at column 10 as specified by the column pointer control ❶**. The *absences* variable will then be read from the next two columns since the length of the informat is 2. SAS will continue reading the rest of the variables as specified by the INPUT statement.

1.3.2   Line Pointer Control (/ or #n)

Syntax:
INPUT <variable> $ <column start – column end>
        / <variable> <column start – column end> ;

or

INPUT #<line number> <variable> $ <column start – column end>
#<line number> variable <column start – column end>;

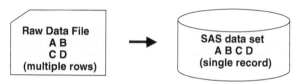

The line pointer control is **used to combine** multiple raw data rows into a single data set row. The line pointer control forces the pointer to a new line of data. In the case of the slash (/), the pointer will go to the next line of data. If the pound-n (#n) line pointer control is used, then the pointer will go to the row specified by the n. **Note that the slash (/) is specified after values of the current record are read, while the pound-n (#n) control is specified before the variable names are defined.** In both cases, when the line pointer control is used, SAS will load the next record into the input buffer and position the column pointer to column one. The pound-n (#n) line pointer control will also allow the skipping of records when reading raw data. Also, it is possible to go back to read skipped records. Note that when using the line pointer control to create single records from multiple raw records, the input file must contain the same number of records for each single record created.

Example 1.7   Read data using the line pointer control.

```
data quiz1;
                         ❶
  input name $ 1-8 / absences 1-2 quiz 3-6;

/* Column Reference Number
12345678901234567890123456789 */
cards;
tim lou
1 6.0
sally
2 10.1
;
run;
data quiz2;
             ❷                    ❸
  input #1 name $ 1-8 #2 absences 1-2 quiz 3-6;
/* Column Reference Number
```

```
1234567890123456789012345567890 */
cards;
tim lou
1 6.0
sally
2 10.1
;
run;
proc print data=quiz2;
run;
```

Output

```
The SAS System

Obs          name    absences      quiz


 1          tim lou    1            6.0
 2          sally      2           10.1
```

In Example 1.7, the line pointer control slash (/) ❶ is used to go to the second line of data after the *name* variable is read. In the quiz2 data set, the pound-*n* (#1) line pointer control ❷ is not required as the column control pointer will start reading data at column 1 of the first line of raw data. The pound-*n* (#2) line pointer control ❸ causes SAS to go to the second line of raw data. Even though we are only displaying quiz1, both quiz1 and quiz2 data sets contain the same information. Note that two records were created from four raw data rows.

1.3.3 Trailing @ Control

Syntax:
INPUT <variable> $ <column start – column end> @;
    IF <variable> IN ('text1','text2');
INPUT <variable1> <variable2> @; OUTPUT;
INPUT <variable1> <variable2>; OUTPUT;

The trailing @ control is used to **read partial raw data or to read an observation more than once** from the raw data. The trailing @ control is generally used with multiple INPUT statements and with an IF statement. The trailing @ control instructs SAS to **hold the current data row for multiple reads.** An IF statement is then used to determine the test condition of a specific variable for further reads, after which the conditional INPUT statement reads the selected data records. The last INPUT statement does not need the trailing @ control; if one exists, SAS ignores it. **This component enables the creation of multiple records from streaming data.** In addition, the trailing @ control prevents, by default, reading multiple raw data rows when using multiple INPUT statements. This is accomplished by storing all data values of the current raw data line in the input buffer for the next INPUT statement to process.

The trailing @ control can be used in the LIST, COLUMN, or FORMAT-TED INPUT statement. All of the data in the current data line are in the input buffer for another INPUT statement to process. If the end of record is encountered on the current data line, a new data line will be loaded into the input buffer. In addition, SAS releases the current data line when the control returns to the top of the DATA step.

Example 1.8   Use multiple INPUT statements with the trailing @ pointer control to read data from the same record more than once.

```
data payment;
  input name $ 1-4 @;                       ❶
    if name in ('bill', 'tom');             ❷
  input pay1 pay2 @; output;                ❸
  input pay1 pay2;     output;              ❹
/* Column Reference Number
12345678901234567890123456789 */
cards;
bill 360 1000 340 2000
john 100 2000 200 3000
;
run;

proc print data=payment;
run;
```

Output

Obs	name	pay1	pay2	
1	bill	360	1000	❺
2	bill	340	2000	

In Example 1.8, SAS will read the first variable and then hold the column position for the next INPUT statement ❶. The IF statement ❷ selects only the records that match the condition; for those selected, SAS reads the remaining data values. The second INPUT statement ❸ reads from the **same data row** but only the next two data values. It then holds its placement again for the next INPUT statement. The OUTPUT statement writes the data values to the data set. The third INPUT statement ❹ reads the remaining two data values in the same two variables as defined in the second INPUT statement. Note that the trailing @ control is not needed in the last INPUT statement, and that two records were created from one raw data row. As a result of the IF condition, records for "john" are excluded from the data set ❺. In this example, the OUTPUT statements are specified to write to two records because the raw data is continuous on the same row. Depending on how the raw data is structured, an OUTPUT statement may not be required when using multiple INPUT statements. Note that an implied OUTPUT statement exists if an OUTPUT statement is not specified.

1.3.4 Double Trailing @@ Control

Syntax:
INPUT <variable1... variablen> @@;

The double trailing @@ control is helpful in creating multiple records in a data set from a single raw line. **The double trailing @@ control instructs SAS to hold the data record until all columns are read.** Thus, SAS holds the record across multiple iterations of the DATA step until the end of the record is reached. As with the trailing @ control, all data values of the current raw data line are stored in the input buffer for the next INPUT statement to process. The number of variables specified in the INPUT statement determines the starting and ending columns.

The double trailing @@ control can be used in the LIST or FORMATTED INPUT statement but should never be used with the COLUMN INPUT statement. This is because with the COLUMN INPUT statement, SAS automatically positions the column pointer as it reads data across the raw data line. The double trailing @@ control should also not be used with the column pointer control @ or the MISSOVER option for the same reason.

Example 1.9   Read data using the double trailing @@ control.

```
data version;
*single input statement for all columns-continuous data
with fixed layout;
input v1 v2 @@;
cards;
20 360 21 240 22 340
;
run;

proc print data=version;
run;
```

Output

```
The SAS System


Obs        v1         v2


 1         20         360   ❶
 2         21         240
 3         22         340
```

In Example 1.9, the INPUT statement has two variables in which data will be read. When the double trailing @@ control is used, all records will be created from one raw data line ❶.

1.3.5   Using the INFILE Statement Options to Control Processing When Reading Raw Data Files

Syntax:
DATA <name>;
  INFILE '<physical file location>' <options>;
  INPUT;
RUN;

The INFILE statement identifies the external file that contains the data instead of having the data in-stream. In-stream data is data that is located in the DATA step after the CARDS or DATALINES statement. Along with specifying the external file, there are several options that can be used with the INFILE statement to control how the data is read. Many of these options have a combined effect. **For example, if both FIRSTOBS=10 and OBS=100 are**

specified, then observations 10 to 100 will be read. A listing of selected INFILE statement options can be found after the example.

In addition, the FILENAME statement can be used with the INFILE statement to reference the external data file. The FILENAME statement assigns a name to the external data file so that only the name needs to be specified in the INFILE statement. This is advantageous if the input file is located elsewhere or if it is passed in as a macro variable. **The INFILE statement can be used with any of the INPUT statements (LIST, COLUMN, FORMATTED, or NAMED).**

Example 1.10   Read formatted input data using the INFILE statement.

```
c:\mydata\data.txt
12345678901234567890123456789  0
tim lou  1     6.0    3 10/23/78
sally    2    10.1    2 01/02/78


filename datfil 'c:\mydata\data.txt';
--------------------------------------------------
data class;
                    ❶        ❷        ❸          ❹
infile datfil dlm=' ' end=last firstobs=2 missover
     ❺
truncover;
   input name $7. @10 absences 2. @15 quiz 4.1 @21 class
   1. @23 dob mmddyy8.;

   if last then put 'Last observation read';
run;


proc print data=class;
run;
```

Output

```
The SAS System

Obs    name      absences    quiz    class    dob

 1     tim lou       1        6.0      3      6870
 2     sally         2       10.1      2      6576
```

In Example 1.10, the data is stored in the data.txt file instead of being listed after the CARDS statement. The FILENAME statement assigns a FILEREF reference that the INFILE statement uses to find the data. Instead of using the FILENAME statement, the INFILE statement can use the full path as an alternative. If Example 1.10 were to only use the INFILE statement without a FILENAME statement it would be as follows:

```
infile 'c:\mydata\data.txt' dlm=' ' end=last firstobs=
2 missover truncover;
```

Table 1.2 lists selected options for the INFILE statement. The bold options with a number correspond to the options used in Example 1.10. In Table 1.2, there are a few options identified by the asterisk (*) that are also data set options. They have the same functionality as when they are used in the INFILE statement and can be applied anywhere a data set name is specified.

In Example 1.11, the DSD option ❶ is used with the DLM option ❷ to treat consecutive commas as a missing value. The *class* variable is missing in the first observation, so when SAS comes to the double commas and the DSD option is specified, SAS will treat the two commas as a missing value for that variable. Without the DSD option, by default SAS will read the date value 10/23/78 into the variable class. This will cause an error because an informat is required to read dates. The DSD option will treat two consecutive delimiters as a missing value and also remove quotation marks from character values.

Example 1.11   Read formatted input data using the INFILE statement with the DSD option.

c:\mydata\data.txt

```
tim lou,1,6.0,,10/23/78
sally,2,10.1,2,01/02/78
--------------------------------------------------
filename datfil 'c:\mydata\data.txt';

data class;
                      ❶      ❷
   infile datfil DSD   dlm=',' ;
   input name $ absences quiz class dob mmddyy8.;
run;


proc print data=class;
run;
```

Output

```
The SAS System

Obs        name          absences        quiz        class        dob
1        tim lou            1            6.0            .         6870
2         sally             2           10.1            2         6576
```

Table 1.2   Selected INFILE Statement Options

INFILE Statement Options	Description/Example
COLUMN = var	A numeric variable that the INPUT statement sets to the column pointer location.
❶ DELIMITERS = 'chars' or var	Delimiters used in the INPUT statement. Default is DLM = ' '; Set DLM = ',' to read comma delimited files, and DLM = '09'x to read tab delimited text files. When reading tab delimited text files created from Excel, it is best to read each variable and not skip variables using the COLUMN INPUT statement.
* ❷ END = var	A numeric variable that the INPUT statement sets to 1 when it reads the last record in the file, unless the UNBUFFERED option is set. Example: END = last; if last then output;
EOF = statement label	An INPUT statement branches to this statement label when it attempts to read past the end of the file.
* ❸ FIRSTOBS = number	The first record to be read from the raw data file. Example; if FIRSTOBS = 10, then first record read is obs # 10.
LENGTH = var	A numeric variable that contains the length of the INFILE_string.
LINE = var	A numeric variable that the INPUT statement sets to the line pointer location.
LINESIZE =	The maximum number of characters used in a record.
LRECL =	Logical record length: the number of characters in a record. Example: LRECL = 1000. If this value is too short, the whole record will not be read. In some operating environments, SAS assumes the external file to have a logical record length of 256 or less. Check your system-specific SAS documentation.
❹ MISSOVER ❺ TRUNCOVER FLOWOVER STOPOVER	Action taken when SAS <u>reaches the end of a record</u> <u>*before* it finds data values for all listed variables in</u> <u>the INPUT statement (i.e., when the raw data row</u> <u>has fewer values than variables).</u>

(continued)

Table 1.2   Selected INFILE Statement Options (*continued*)

INFILE Statement Options	*Description/Example*
	MISSOVER: assigns missing values to the remaining variables without starting to read the next data line. TRUNCOVER: prevents loading another record. FLOWOVER: continues reading at column 1 of the next record. STOPOVER: creates an error condition and stops the DATA step. Thus, by specifying both MISSOVER and TRUNCOVER options, SAS automatically assigns missing values if there are fewer raw data values than variables in the INPUT statement.
N =	The number of lines available to the line pointer control.
* OBS =	The last record to be read from the raw file. Example: OBS = 100.
PAD/NOPAD =	Pads short records with trailing blanks. This is the default for fixed-length records.
RECFM =	Record format length. Options are fixed, variable, undefined, data sensitive, and no format.
START = var	A numeric variable whose value indicates the first character to be used in the_infile string.
UNBUFFERED =	Does not look ahead at the next record when reading a record.
DSD	In the INPUT statement, allows for reading comma delimited or quoted string files. Sets the default delimiter to a comma. Treats two consecutive delimiters as a missing value and removes quotation marks from character values. Without the DSD option, each delimiter is not treated as a separator because consecutive delimiters are treated as one separator.

*Also works as a data set option.

1.4   Importing Data from Excel and Access Using the IMPORT Procedure

With SAS, it is easy to import data from Excel or Access using the IMPORT procedure. The approach shown in this section reads a comma delimited text file created from Excel and Access. The IMPORT wizard can be used to automate this process and automatically create the SAS code.

To import data directly, without the comma delimited text file, from Excel or Access files requires the SAS/ACCESS to PC module. The only change required would be to specify the Excel file name in the DATAFILE = option and to include the SHEET = option with the sheet name. Directly importing data from Access tables, without the tab delimited text file, requires the following: TABLE = Access table name, DATABASE = database name, DBMS = access, USID = userid, and PWD = password.

Syntax:
PROC IMPORT OUT= \<data set name\>
   DATAFILE= "\<physical file location\>"
   DBMS=\<CSV I TAB\> \<REPLACE\>;
   GETNAMES=\<YES I NO\>;
   DATAROW=\<row number\>;
RUN;

Example 1.12   Read an Excel file using a comma separated file.

```
Comma Separated File: class.csv ❶
name,absences,quiz,class,weight ❷
tim,1,6,3,  ❸
sally,2,10,4,110
john,-2,8,3,120
--------------------------------------------------

proc import out= work.class
        datafile= "c:\mydata\class.csv"
        dbms=csv replace;                      ❹
        getnames=yes;
        datarow=2;
run;

proc print data=class;
run;

proc contents data=class;
run;
```

Output
```
The SAS System
```

Obs	name	absences	quiz	class	weight
1	tim	1	6	3	.
2	sally	2	10	4	110
3	john	-2	8	3	120

Contents

```
-----Alphabetic List of Variables and Attributes-----

   #   Variable    Type   Len   Pos   Format   Informat

   2   absences    Num     8     0    BEST12.   BEST32.
   4   class       Num     8    16    BEST12.   BEST32.
   1   name        Char    6    32       $6.       $6.
   3   quiz        Num     8     8    BEST12.   BEST32.
   5   weight      Num     8    24    BEST12.   BEST32.
```

❺

The first step in Example 1.12 is to save the Excel file as a comma separated file, class.csv ❶. The Excel file has three records and five variables. Since the variable names are in the first row ❷, GETNAMES = YES and DATAROW = 2 options are specified. The data set created contains all values in the Excel file ❸. The IMPORT procedure will read the comma separated file, class.csv, to create the class data set since the DBMS = CSV is specified ❹. The REPLACE option instructs SAS to replace the data set if it already exists. You can use the IMPORT wizard to create this code automatically. As you can see, SAS correctly created the variables, automatically determined the variable type and assigned the default lengths ❺. As an alternative to the IMPORT procedure, the LIST INPUT statement could also read the raw data file. See Example 1.1 for more information.

Example 1.13   Read an Access table using a tab delimited file.

```
Tab Delimited File: class.txt ❶
"name"     "absences"        "quiz"       "class"  "weight" ❷
"tim"        1.00             6.00          3.00     "."    ❸
"sally"      2.00            10.00          4.00     "110"
"john"      -2.00             8.00          3.00     "120"
---------------------------------------------------------
proc import out= work.class2
      datafile= "c:\mydata\class.txt"
      dbms=tab replace;                    ❹
      getnames=yes;
      datarow=2;
run;


proc contents data=class2;
run;
```

```
proc print data=class2;
run;
```

Output

```
The SAS System

Obs       name      absences      quiz      class     weight

 1        tim          1            6          3          .
 2        sally        2           10          4         110
 3        john        -2            8          3         120
```

Contents

```
-----Alphabetic List of Variables and Attributes-----
       #   Variable   Type   Len   Pos    Format   Informat

❺      2   absences   Num     8     0     BEST12.   BEST32.
       4   class      Num     8    16     BEST12.   BEST32.
       1   name       Char    6    32        $6.       $6.
       3   quiz       Num     8     8     BEST12.   BEST32.
       5   weight     Num     8    24     BEST12.   BEST32.
```

The first step in Example 1.13 is to save the Access table as a comma separated file or a tab delimited file, such as class.txt ❶. The Access table has three records and five variables. Since the variable names are in the first row ❷, GETNAMES = YES and DATAROW = 2 options are specified. The data set created contains all records in the Access table ❸. The IMPORT procedure will read the tab separated file, class.txt, to create the class2 data set since the DBMS = TAB is specified ❹. You can use the IMPORT wizard to create this code automatically. As you can see, SAS correctly created the variables, automatically determined the variable type, and assigned the default lengths ❺.

1.5   SET Statement to Read and Combine SAS Data Sets

There are multiple ways in which to combine two or more SAS data sets. In this section, the SET statement with and without the BY statement is used to combine data sets.

1.5.1 General Rules

The SET statement is used to access SAS data sets. In general, SAS reads observations from the input data set sequentially unless the POINT = option is specified to access observations directly. If using the POINT = option, make sure to specify the OUTPUT and STOP statements to write to the output data set and to prevent continuous looping. In addition, the END = option cannot be used with the POINT = option.

The SET statement can also be used to combine two or more data sets together. The combined data set will contain variables from all of the input data sets and will be the concatenation of all the records from the input data sets. Common variables in the data sets must have matching data types. In addition, **the length of common variables will be defined from the first data set in the SET statement. The FORMAT, INFORMAT and LABELs of common variables will also be defined from the first data set in the SET statement. Note that if the FORMAT, INFORMAT or LABEL statement is not specified in the first data set, than SAS uses the FORMAT, INFORMAT and LABEL statement from the second data set if specified.** Note also that this rule is the same when using the MERGE Statement but different when using the SQL procedure. For uncommon variables, missing values will be assigned to the variables from which with the data set the variables do not exist. If the variable is numeric the variable will be set to . and blank for the character variables. Table 1.3 lists the two types of SET statements used to combine data sets. In general, data sets combined using the SET statement should contain the same variables.

Table 1.3   Two Methods of Using SET Statements to Combine Data Sets

Type	Concatenation	Interleaving
BY statement	No	Yes
Process	Appends records	Appends and sorts records

1.5.2 SET Statement

Syntax:
SET <data set1... data setn>;

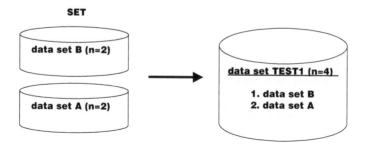

When combining data sets using the SET statement without the BY statement, the observations are concatenated, i.e., SAS places observations from the second data set after the observations from the first data set. The combined data set is the sum of observations from all data sets. In addition to combining the data sets, new variables can be created and the data set can be subsetted.

Example 1.14   Read data sets using the SET statement.

```
data A;
  input patno source $ gender $;
  cards;
1 A male
3 A male
;
run;

data B;
  input patno source $ age;
  cards;
2 B 45
4 B 35
;
run;

data test1;
  set B A; ❷
run;

proc print data=test1;
run;
```

Annotated Output

```
The SAS System              ❸
      ❶ ( Data A, B )    (Data A)(Data B)
Obs   patno     source   age  gender
 1      2         B       45            (data set B obs, gender
                                        set to missing)
 2      4         B       35            (data set B obs, gender
                                        set to missing)
 3      1         A        .    male    (data set A obs, age set
                                        to missing)
 4      3         A        .    male    (data set A obs, age set
                                        to missing)
```

In Example 1.14, both of the data sets contain two observations, so the combined data set is the concatenation of the two data sets or four observations. The common variables *patno* and *source* have data values in the same corresponding variables in the test1 data set. ❶ Note that data set B records are before data set A records because data set B was specified as the first data set in the SET statement. ❷ In addition, the *gender* and *age* variables have missing values for the observations where they did not exist in the original corresponding data set A and B. ❸

1.5.3   SET Statement with the BY Statement

Syntax:
SET <data set1...data setn>;
BY <variable>;

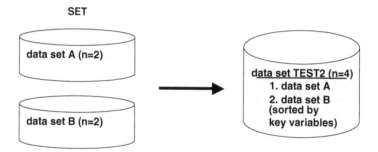

BY Key Variables

Interleaving SAS data sets is done using the BY statement with the SET statement. This requires all BY variables to match in variable name, type, and attributes. The new combined data set will be sorted in the order of the BY variables. Both of the data sets **must be sorted by the BY variables** in advance of the SET statement or be indexed by the BY variables. If the variables in the BY statement need to be sorted in descending order, then the keyword DESCENDING will also be used before the variables' names. Note that if the BY variables do not have the same name in both data sets, then the BY variable name in one data set can be changed using the RENAME = data set option that will be introduced in Chapter 2, Section 2.2.

   When the BY statement is used, two additional variables, FIRST. by_variable and LAST.by_variable, are created for each BY variable. These variables are temporary variables that exist only in the DATA step and flag changes in the BY variables' values. For similar data values in the

BY variables, **the FIRST.by_variable and LAST.by_variable represent the first and last occurrences of that variable's value.** For the first record, FIRST.by_variable will equal 1 and 0 for all other records. For the last record, LAST.by_variable will equal 1 and 0 for all other records. Thus, the FIRST.by_variable and LAST.by_variable can be used as conditions for actions that need to be done only once for each group of the BY variable's value. **Note that when specifying multiple BY variables, a change in the value of the first BY variable forces the LAST.by_variable to equal 1 for the other BY variables.**

Example 1.15   Combine data sets using the SET and BY statements.

```
data A;
  input patno source $ gender $;
  cards;
1 A male
3 A male
;
run;

data B;
  input patno source $ age;
  cards;
2 B 45
4 B 35
;
run;

proc sort data=A;
  by source;
run;

proc sort data=B;
  by source;
run;

data test2;
    set A B;
    by source;
```

```
/* Condition                    Execute the statement at the */
/* If first.source = 1 then;  Beginning of the group */
/* If last.source = 1 then;    End of the group */


run;


proc print data=test2;
run;
```

Annotated Output

```
The SAS System
          ( Data A, B )   (Temp Variables)              (Data A) (Data B)
  Obs    patno   source    first.source   last.source    gender    age

   1       1       A          1    ❶         0            male       .
   2       3       A          0              1   ❷        male       .
   3       2       B          1              0                       45
   4       4       B          0              1                       35
```

In Example 1.15, the observations are sorted by the *source* variable. As in Example 1.14, the number of observations is the sum of the observations from the input data sets, and missing values are created for observations that do not exist in the original data set A and B. Note that in the annotated output, the variables *first.source* and *last.source* are not saved with the data set and are just displayed to show how the values change. The *first.source* variable equals 1 for the first record of each new value of the *source* variable ❶. The *last.source* variable equals 1 for the last record of each new value of the *source* variable ❷. The value for all other records are set to zero. As you can see, the temporary variables can be used to execute SAS statements at the beginning or at the end of the group.

1.6   Using the MERGE Statement to Combine SAS Data Sets

There are multiple ways in which to combine two or more SAS data sets. In this section, the MERGE statement with and without the BY statement is used to combine data sets.

1.6.1   General Rules

There are several types of MERGE statements that will be discussed, all of which follow the same rules as the SET statement. All of the variables that are in common within the data sets must be of the same type. Also, for common variables, the **length will be defined from the first data set in the MERGE statement. The FORMAT, INFORMAT and LABELs of common variables**

Table 1.4   Two Types of Merges

Type 1 Merge: One-to-One Does Not Use BY Statement*	Type 2 Merge: Match-Merge Uses BY Statement
Records are combined with records	Matching BY variables are combined
First observation in one data set with first observation from another data set	Only one BY statement
	Each data set must have the BY variable
If number of observations or variables are not the same between the data sets, then missing values are assigned accordingly.	

*Not recommended as this can cause linking of unrelated records.

will also be defined from the first data set in the MERGE statement. Note that if the FORMAT, INFORMAT or LABEL statement is not specified in the first data set, then SAS uses the FORMAT, INFORMAT and LABEL statement from the second data set if specified. As discussed in the previous section, this is the same rule when using the SET statement but different when using the SQL procedure. One difference from the SET statement is that with the MERGE statement, the common variables from the **second data set** will overwrite the data values from the first data set. Note that this is a different rule than with the SQL procedure. This only occurs when you have matching values for the variables in the BY statement. When you have unmatching values, the values from each data set are maintained. Also, as with the SET statement, uncommon variables will be set to missing for corresponding records that are in the other data set. Table 1.4 shows the two types of merges.

1.6.2   Type 1 Merge: One-to-One

Syntax:
MERGE <data set1...data setn>;

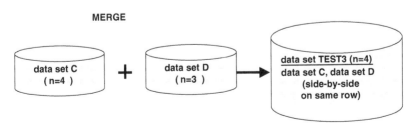

The first type of merge is the one-to-one merge in which only the MERGE statement is used in the DATA step without a BY statement. In a one-to-one merge, the number of observations in the final data set will be equal to the **number of observations from the largest input data set.** SAS will pair the observations from each of the data sets, one record to one record. Another way to accomplish this is to specify the SET statement for each data set. Since there is no BY statement, the data sets will not be merged by BY variables. In general, merging data sets together without a BY statement is not recommended because the records in each data set are joined only by their positions within the data sets and not by a common unique key variable. **Almost all MERGE statements will be followed by a BY statement to link related information.**

Example 1.16   Performing one-to-one merge using the MERGE statement.

```
data C;
  input patno source $ gender $;
  cards;
1  C  male
2  C  female
3  C  male
5  C  male
;
run;

data D;
  input patno source $ age;
  cards;
2  D  45
2  D  40
4  D  35
;
run;

data test3;
   merge C D;
run;

proc print data=test3;
run;
```

Annotated Output

```
The SAS System
      (  Data C, D  )     (Data C)(Data D)
```

Obs	patno	❶ source	gender	age	
1	2	D	male	45	(data sets C & D, records paired)
2	2	D	female	40	(D values for patno, source)
3	4	D	male	35	(D values for patno, source)
4	5	C	male	.	(Extra data set C obs, age from data set D set to missing)

In Example 1.16, the first three records from data set C are paired with the three records from data set D ❶. Data values for the common variables are saved from the second data set, data set D. The unpaired fourth record from data set C is kept in the new data set with the unique variables from data set D set to missing. The total number of records is equal to the data set with the most records, which is data set C with four records.

Example 1.17 Performing one-to-one merge using multiple SET statements.

```
data C;
  input patno source $ gender $;
  cards;
1 C male
2 C female
3 C male
5 C male
;
run;

data D;
  input patno source $ age;
  cards;
2 D 45
2 D 40
4 D 35
;
run;

data test3;
   set C;
   set D;
run;
```

```
proc print data=test3;
run;
```

Annotated Output

```
The SAS System
         (  Data C,  D  )      (Data C)(Data D)
 Obs    patno  ❶  source   gender    age

  1       2          D        male     45    (data sets C & D, records
                                             paired)
  2       2          D        female   40    (D values for patno,
                                             source)
  3       4          D        male     35    (D values for patno,
                                             source)
```

Similar to performing a one-to-one merge with the MERGE statement is using multiple SET statements. In Example 1.17, the first three records from data set C are paired with the three records from data set D ❶. Data values for the common variables are saved from the second data set, data set D. **With the multiple SET statements, however, the unpaired fourth record from data set C is not kept in the new data set because data set D has only three records.** The last record number processed by SAS is from data set D because it has fewer records than data set C. The fourth record from data set C is not processed. **The total number of records in the new data set is equal to the number of records in the smallest data set, which is data set D with three records.** In general, it is not recommended that you use multiple SET statements to merge data sets unless this is required. A better alternative is to use the MERGE statement without the BY statement to perform the one-to-one merge.

1.6.3   Type 2 Merge: Match Merge

Syntax:
MERGE <data set1...data setn>;
BY <variable>;

BY Key Variables

The difference between the type 1 and type 2 (one-to-one and match) merges is that the type 2 (match) merge uses the BY statement. If the BY statement is used, SAS will pair the observations only when BY variables match. This requires all BY variables to match in variable name, type, and attributes. New observations will be created when the BY variables do not match. The new data set will then be a combination of all the **unique observations for the BY variables. The match merge requires the data sets to be presorted or indexed** by the BY variables before the MERGE statement. The output data set will be sorted in the order of the BY variables. If the data set is sorted in descending order, the DECENDING keyword can be used before the variable's name to indicate the BY variables are sorted in descending order. Note that if the BY variables do not have the same name in both data sets, then the BY variable name in one data set can be changed using the RENAME = data set option that will be introduced in Chapter 2, Section 2.2.

To help control and eliminate non-matches in the new data set, a record contributor variable can be used. The data set option (IN=) creates temporary variables that can be used with the IF statement to subset the data set; e.g., MERGE C D (IN=IND); BY PATNO; IF IND;. The final data set will only contain records that were in the original D data set.

Example 1.18   Performing a match merge using the MERGE and BY statements.

```
data C;
  input patno source $ gender $;
  cards;
1  C  male
2  C  female
3  C  male
5  C  male
;
run;

data D;
  input patno source $ age;
  cards;
2  D  45
2  D  40
4  D  35
;
run;
```

```
proc sort data=C;  ❶
  by patno;
run;

proc sort data=D;
  by patno;
run;

data test4;
      merge C D;  ❷
      by patno;
run;

proc print data=test4;
run;
```

Annotated Output

```
The SAS System
          ( Data C, D )     (Data C) (Data D)
  Obs     patno    source   gender    age
```

Obs	patno	source	gender	age	
1 ❸	1	C	male	.	(data set C, age set to missing)
2	2	D	female	45	(2 obs due to 2 obs in data set D)
3	2	D	female	40	(D values for patno and source)
4	3	C	male	.	(data set C, age set to missing)
5	4	D		35	(data set D, gender set to missing)
6	5	C	male	.	(data set C, age set to missing)

In Example 1.18, data sets are first sorted by the BY variable *patno*.❶ Using the MERGE with the BY statement ❷ produces the combined data set with 6 observations ❸. Observations 1, 4, and 6 are unique to data set C, so the variable (*age*) from data set D is set to missing. Observations 2 and 3 have values for the *patno* variable that match in both data sets C and D. As before, data values for common variables are saved from the second data set, so the values of the *source* variable will be set to D. In observation 3, the value of the *gender* variable from data set C is retained from observation 2 because there are two records where *patno* = 2 in data set D. Observation 5 is unique to data set D, so the unique variables from data set C (gender variable) will be set to missing.

Example 1.19   Performing a match merge using MERGE statement with the IN= option and the BY statement.

```
data C;
  input patno source $ gender $;
  cards;
1 C male
2 C female
3 C male
5 C male
;
run;

data D;
  input patno source $ age;
  cards;
2 D 45
2 D 40
4 D 35
;
run;

proc sort data=C;
  by patno;
run;

proc sort data=D;
  by patno;
run;

data test5;
    merge C (in=inc)  D (in=ind); ❶
    by patno;
    if inc and ind;                    ❸
run;

proc print data=test5;
run;
```

Annotated Output

Obs	patno	source	gender	age	inc ❷	ind
1	2	D	female	45	1	1
2	2	D	female	40	1	1

The only difference in the MERGE statement between Example 1.18 from Example 1.19 is the addition of the data set option (IN=) for both data sets. ❶ The data set option (IN=) creates the temporary variables (*inc*, *ind*) that can be used with the IF statement to subset the data. In this example, all of the data from the data set D will have the temporary variable *ind* set to 1 while the data from data set C will have the temporary variable *inc* set to 1. ❷ When the IF *inc* and *ind* condition is executed, only the data that has common values for the *patno* variable from data sets C and D will be kept. ❸

1.7 SQL Procedure to Create Data Sets

The Structured Query Language (SQL) procedure is useful in that it can do all of the things that the SET and MERGE statements can do plus more. **The SQL procedure, in addition, can sort data, add or modify columns in a table, add or modify values in a table, create tables, generate reports and retrieve and manipulate SAS tables.** In the SQL procedure, the term table means the same thing as data set. Likewise, columns and rows in the SQL procedure are the same as variables and records in the data set. Many of the SQL procedure examples in this section show the corresponding DATA step statements to accomplish the same task. Note that results may differ for common variables when joining data sets as compared to merging data sets. In general, you will find that fewer SQL procedure statements are needed than DATA step statements for the same task.

1.7.1 SQL Procedure

Syntax:
PROC SQL;
  CREATE TABLE <data set> **AS**
  SELECT <variable1,...,variablen>, <variable expression> **AS** <variable>
  FROM <data set>
  WHERE <variable>='text'
  ORDER BY <variable>;
QUIT;

The SQL procedure is different from other SAS procedures in that it uses clauses (WHERE, FROM, and ORDER BY) instead of just SAS statements. The general form of the SQL procedure starts with the PROC SQL statement, which invokes the SQL procedure. If you are creating a data set, then the CREATE TABLE AS clause goes after the PROC SQL statement. The next clause is the SELECT clause, which specifies which columns will be selected. The SELECT keyword is followed by the list of column names separated by commas. The AS keyword can be used in the SELECT statement to create a new column that will be assigned the value of the preceding column or expression. The FROM clause is used to specify the tables that are to be queried. The tables are listed after the FROM keyword separated by commas. Following the FROM clause is the optional WHERE clause. The WHERE clause subsets the data based on a specified condition. The expression that follows the WHERE keyword can be any valid SAS expression. The last statement is the ORDER BY clause. The ORDER BY clause sorts the rows by the values of the specified columns. An additional statement that can be used is the GROUP BY clause. The GROUP BY clause is used in queries that include one or more summary functions. The GROUP BY clause will produce a statistical summary for each group that is defined in the GROUP BY statement. To complete the PROC SQL statement, a QUIT statement is required. This is different from most SAS procedures since no RUN statement is needed to end the procedure. Example 1.20 shows how the SQL procedure is used to create a data set.

Example 1.20   Using the SQL procedure to create data sets.

```
data C;
  input patno source $ gender $;
  cards;
1 C male
2 C female
3 C male
5 C male
;
run;
```

```
proc sql;                             /* Comparable DATA Step
                                         and PROC SORT */
  create table test6 as ❶         /* DATA TEST6; */
  select patno, source, gender,   /*    KEEP PATNO SOURCE
                                         GENDER WEIGHT; */
  150 as weight ❷                 /*       WEIGHT = 150; */
  from C                          /*       SET C; */
  where gender='male' ❸      /*     WHERE GENDER =
                                         'male'; */
```

```
    order by patno;  ❹        /* BY  PATNO;  RUN;
                                 PROC  SORT  DATA = TEST6 BY
                                 PATNO;*/
quit;                                    /*  RUN;  */


proc print data=test6;
run;
```

Annotated Output

```
The SAS System
         (   data set  C  )
 Obs    patno   source   gender   weight

  1       1       C       male      150
  2       3       C       male      150
  2       5       C       male      150
```

In Example 1.20, you can see how the SQL procedure compares to the DATA step and the SORT procedure. One of the biggest advantages of using the SQL procedure is being able to do these steps within one procedure. The SQL procedure selects the variables that will be kept in the final data set, ❶ creates any new variable, ❷ subsets the data based on the WHERE clause ❸ and orders the records. ❹ Data sets need not be presorted for merging using the SQL procedure.

1.8 SQL Procedure to Query Multiple Tables

1.8.1 *General Rules*

The SQL procedure allows you to combine up to 16 tables at a time. In Version 9.1, SAS allows up to 32 tables to be joined. See Chapter 6, Section 6.3 for more information about Version 9.1 upgrades. When using the SQL procedure to combine multiple tables, there are several different types of joins that are available. In this section, the two different types of joins, the INNER and the OUTER join, are used to combine tables. The INNER join combines records that are common between two data sets. With the OUTER join, there are three different types: the LEFT, RIGHT, and FULL join. The diagrams below show which data set, left or right, will be saved in the final data set. The shaded data set side will have all records in the final data set.

All the joins have several general rules. **When combining data sets with common variables, the LENGTH, FORMAT, INFORMAT and LABELS for**

OUTER JOIN: **LEFT**	**RIGHT** **FULL**

the common variables are defined from the first or left data set and not from the second or right data set. For the FORMAT, INFORMAT and LABELS, this rule is different when using the DATA step to combine data sets. In addition, when using the SQL procedure, data values from the first or left data set are saved and not replaced by the values from the second data set. Note that this is the opposite rule to using the MERGE statement in the DATA step. To prevent SAS from replacing data values, instead of using the asterisk to combine all variables from both data sets, select the common variable from only one of the data sets and list all of the other variables required. See Example 1.26 to select individual variables from each data set.

Note also that when joining data sets using SQL procedure, SAS will issue a WARNING message if trying to select common variables from data sets without selecting them from individual data sets or using the COALESCE() function. It is recommended to change the selection of variables to prevent this WARNING message. When using the MERGE with the BY statement, however, to perform a similar task, SAS does not issue a WARNING message when selecting common variables from data sets.

1.8.2 INNER Join

Syntax:
PROC SQL;
  CREATE TABLE <data set> AS
  SELECT <alias1>.<variable1>, <alias2>.<variable2>
  FROM <data set1> AS <alias1>, <data set2> AS <alias2>
  WHERE <alias1>.<variable>=<alias2>.<variable>;
QUIT;

The INNER join will return a result table for all of the rows in a table that have one or more matching rows in another table. This will give the same result as using the MERGE with a BY statement and then condition to select only matching observations. When using the INNER join, the maximum number of tables or views that can be combined is 16.

Example 1.21 **Perform INNER join between two data sets.**

```
data C;
  input patno source $ gender $;
  cards;
1 C male
2 C female
3 C male
```

```
5 C male

;

run;


data D;
  input patno source $ age;
  cards;
2  D  45
2  D  40
4  D  35

;

run;



proc sql;                        /* Comparable DATA Step except
                                   for common variables*/
  create table test7 as          /* DATA TEST7; */
  select *
  from C as c,        /*     MERGE C (IN=INC) D (IN=IND); */
  D as d
  where c.patno=d.patno;  /*   BY PATNO; IF INC AND IND; */
quit;                            /* RUN; */


proc print data=test7;
run;
```

Annotated Output

```
The SAS System
        (    Data C, D ) (Data C)  (Data D)
Obs   patno    source   gender   age

 1      2        C      female   45   (Patno = 2 exists in
                                       both data sets)
 2      2        C      female   40   (C values for patno
                                       and source)
```

In Example 1.21, you can compare the SQL procedure to the DATA step. With the DATA step, both input data sets would have to be presorted by the BY variable *patno*. With the SQL procedure, you do not have to sort the variables in advance. Using the same data as before, the results are two observations in which the *patno* variable is in both data sets. This is comparable to the data set created in Example 1.19 except for the values in the common variable *source*. With the SQL procedure, the *source* variable is set to C because it was the first data set instead of being set to D for when using a DATA step.

When combining data sets with the INNER join it is important to use a WHERE clause. **If an INNER join is done without the WHERE clause the resulting table will be a CARTESIAN PRODUCT.** Therefore, the resulting table will be all possible combinations of each record between the input data sets. **This can cause very large and unexpected data sets.** Example 1.22 shows the combination of our data sets without the WHERE clause. Note that there is not a comparable DATA step for performing an INNER join without a WHERE clause.

Example 1.22   Perform an INNER join without a WHERE clause.

```
data C;
  input patno source $ gender $;
  cards;
1 C male
2 C female
3 C male
5 C male
;
run;

data D;
  input patno source $ age;
  cards;
2 D 45
2 D 40
4 D 35
;
run;

proc sql;
  create table test8 as
  select *
  from C as c,
  D as d      /* without a WHERE clause */
  order by patno;
quit;

proc print data=test8;
run;
```

Annotated Output

```
The SAS System
         (   Data  C,  D   )(Data  C)  (Data  D)(Cartesian  Product)
  Obs    patno    source   gender  age

   1      1          C      male     45 (data set C, obs=1, D, obs=1)
   2      1          C      male     40          data set D, obs=2)
   3      1          C      male     35          (data set D, obs=3)
   4      2          C      female   45 (data set C, obs=2, D, obs=1)
   5      2          C      female   40          (data set D, obs=2)
   6      2          C      female   35          (data set D, obs=3)
   7      3          C      male     45 (data set C, obs=3, D, obs=1)
   8      3          C      male     40          (data set D, obs=2)
   9      3          C      male     35          (data set D, obs=3)
  10      5          C      male     45 (data set C, obs=4, D, obs=1)
  11      5          C      male     40          (data set D, obs=2)
  12      5          C      male     35          (data set D, obs=3)
```

In Example 1.22, each observation from the first data set is individually paired with each observation from the second data set. The resulting number of observations is the number of observations from the first data set times the number of observations from the second data set; e.g., $4 \times 3 = 12$.

1.8.3 OUTER LEFT Join

Syntax:
PROC SQL;
  CREATE TABLE <data set> AS
  SELECT <alias1>.<variable1>, <alias2>.<variable2>
  FROM <data set1> AS <alias1>,
  LEFT JOIN <data set2> as <alias2>
  ON <alias1>.<variable>=<alias2>.<variable>;
  QUIT;

The OUTER LEFT join is similar to an INNER join except that all of the rows from the first data set are included instead of just the matching rows in the new data set. This is also similar to the MERGE statement with a BY statement and then a condition to select all of the records from one or more of the first data set. The data set that is returned may contain more rows than the number of records in any of the data sets because all combinations of records with duplicate join conditions are retrieved. The OUTER LEFT join uses the ON clause to specify the variables to match between the data sets.

Example 1.23   Perform an OUTER LEFT join between two data sets.

```
data C;
 input patno source $ gender $;
 cards;
1 C male
2 C female
3 C male
5 C male
;
run;

data D;
 input patno source $ age;
 cards;
2 D 45
2 D 40
4 D 35
;
run;

proc sql;              /* Comparable DATA Step except
                              for common variables*/
 create table test9 as  /* DATA TEST9; */
 select *                      ❶              ❷
 from C as c             /*   MERGE C (IN=INC) D (IN=IND); */
 left join D as d        /*   IF INC; */ ❸
 on c.patno=d.patno; /*   BY PATNO; */ ❹
quit;                   /* RUN; */

proc print data=test9;
run;
```

Annotated Output

```
The SAS System
       (Data C, D)        (Data C) (Data D)(All data set C obs)
 Obs    patno    source    gender    age

  1       1        C        male      .    (No matching D obs, age is
                                             missing)
  2       2        C        female    45   (Matching obs from C & D
                                             data sets)
```

3	2	C	female	40	(Matching obs from C & D data sets)
4	3	C	male	.	(No matching D obs, age is missing)
5	5	C	male	.	(No matching D obs, age is missing)

In Example 1.23, the data set C is considered the LEFT table ❶ and the data set D is considered the RIGHT table ❷. Since the LEFT join clause is specified and data set C is considered the LEFT table, all of the records from data set C will be saved ❸. The *patno* variable ❹ is used in the ON clause to match the records between the two data sets. Only matching observations from data set D are saved to the final data set. Note that this is similar to the WHERE clause in Example 1.2.1.

All four records from data set C are saved to the final data set. Only two records are saved from data set D since they have the same *patno* variable values as observations in data set C. For the two records that match, the source variable will be equal to C since the rule is that the variables from the first data set will be kept for common variables.

1.8.4   OUTER RIGHT Join

Syntax:
PROC SQL;
  CREATE TABLE <data set> AS
  SELECT <alias1>.<variable1>, <alias2>.<variable2>
  FROM <data set1> AS <alias1>,
  RIGHT JOIN <data set2> as <alias2>
  ON <alias1>.<variable>=<alias2>.<variable>;
QUIT;

The OUTER RIGHT join has similar syntax to the OUTER LEFT join. The only differences are that LEFT is changed to RIGHT and that all of the observations are kept from the second data set instead of the first data set.

Example 1.24   Perform an OUTER RIGHT join between two data sets.

```
data C;
  input patno source $ gender $;
  cards;
1 C male
2 C female
3 C male
5 C male
;
run;
```

```
data D;
  input patno source $ age;
  cards;
2 D 45
2 D 40
4 D 35
;
run;
```

```
proc sql;                        /* Comparable DATA step except
                                    for common variables*/
   create table test10 as  /* DATA TEST10; */
   select *                      ❶                    ❷
   from C as c                   /*MERGE C (IN=INC) D (IN=IND); */
   right join D as d             /*    IF IND; */ ❸
   on c.patno=d.patno;           /*    BY PATNO; */ ❹
quit;                            /* RUN; */
```

```
proc print data=test10;
run;
```

Annotated Output

The SAS System

	(Data C, D)		(Data C)	(Data D)	(All data set C obs)
Obs	patno	source	gender	age	
1	2	C	female	45	(Matching obs from C & D data sets)
2	2	C	female	40	(Matching obs from C & D data sets)
3	.			35	(No Matching C obs, patno, source and gender are missing)

In Example 1.24, the data set C is considered the LEFT table ❶ and the data set D is considered the RIGHT table ❷. The RIGHT join clause is specified and data set D is considered the RIGHT table, so all of the records from data set D will be saved ❸. The *patno* variable ❹ is used in the ON clause to match the records between the two data sets. Only matching observations from data set C are saved to the final data set.

All three observations from the D data set are saved. For the first two observations there is a matching record in dataset C. Since the SQL procedure takes the values from the first dataset the source variable is set to C. In the last observation there is no observation from dataset C to match with the observation in data set D. Because of this, the values for *patno, source*, and *gender* variables are set to missing but the value of the *age* variable comes from dataset D.

1.8.5   OUTER FULL Join

Syntax:
PROC SQL;
  CREATE TABLE <data set> AS
  SELECT <alias1>.<variable1>, <alias2>.<variable2>
  FROM <data set1> AS <alias1>,
  FULL JOIN <data set2> as <alias2>
  ON <alias1>.<variable>=<alias2>.<variable>;
QUIT;

The OUTER FULL join is also similar to the OUTER RIGHT and LEFT joins except that the FULL join selects all of the matching and nonmatching observations. Thus, records from both data sets are saved in the final data set.

**Example 1.25   Perform an OUTER FULL join between two
                data sets.**

```
data C;
  input patno source $ gender $;
  cards;
1  C  male
2  C  female
3  C  male
5  C  male
;
run;

data D;
  input patno source $ age;
  cards;
2  D  45
2  D  40
4  D  35
;
run;
```

```
proc sql;                        /* Comparable DATA step except
                                    for common variables*/
  create table test11 as    /* DATA TEST11; */
  select *                            ❶                    ❷
  from C as c                    /* MERGE  C  (IN=INC)  D
                                    (IN=IND); */
  full join D as d
on c.patno=d.patno;            /* BY PATNO; */ ❹
quit;                              /* RUN; */

proc print data=test11;
run;
```

Annotated Output

```
The SAS System
          (Data C, D) ❸        (Data C) (Data D)   (All data set C obs)
Obs    patno    source   gender    age

 1       1        C        male       .      (No matching D obs, age
                                              is missing)
 2       2        C        female    45      (Matching obs from C & D
                                              data sets)
 3       2        C        female    40      (Matching obs from C & D
                                              data sets)
 4       3        C        male       .      (No matching D obs, age
                                              is missing)
 5       .                            35     (No matching C obs,
                                              patno, source and gender
                                              are missing)
 6       5        C        male       .      (No matching D obs, age
                                              is missing)
```

In Example 1.25, the result of the FULL JOIN is a match of all the matching and nonmatching observations. Since data set C is the first data set ❶ and data set D is the second data set ❷, variables that have matching observations will take the values of data set C ❸. Since this is a FULL JOIN, all of the observations from both data sets are selected. The *patno* variable ❹ is used in the ON clause to match the records between the two data sets.

Observations 2 and 3 are observations where the values match between the two datasets. Since the SQL procedure takes the values from the first table, the values of the *source* variable are set to C. Observations 1, 4, and 6 are the observations where there is no matching observation in dataset D. For these observations, the value for the *age* variable is missing. Observation 5 has no matching observation in data set C. Since there is no matching observation in the first data set, the values for the *patno, source,* and *gender* variables are set to missing. The value for the age variable will come from data set D. This is comparable to the data set created in Example 1.18 except for the values in the common variables *patno* and *source*. With the SQL procedure, the values from the first data set are selected instead of selecting BY variable values from the second data set when using the MERGE and BY statements within a DATA step.

1.8.6 COALESCE Clause

Syntax:
PROC SQL;
 CREATE TABLE <data set>AS
  SELECT COALESCE (<alias1>.<variable1>,<alias2>.<variable1>) as <variable1>,
       COALESCE (<alias1>.<variable2>,<alias2>.<variable2>) as <variable2>,
       <alias>.<variable>,...,<alias>.<variable>
  FROM <data set1> AS <alias1>.,
  FULL JOIN <data set2> as <alias2>
  ON <alias1>.<variable>=<alias2>.<variable>;
QUIT;

In Example 1.25 (the FULL JOIN) the first data set C did not have a matching observation to the second data set D. The values of the *patno* and *source* variables, which are common to the two data sets, were set to missing. **The COALESCE ( ) function can be used for key variables where the variable's value can come from either of the data sets.** This can be useful if a key variable is missing in one of the data sets, as the value can then come from the other data set. The COALESCE ( ) function will go through each of its arguments until it finds the first nonmissing value.

Example 1.26   **Using the COALESCE ( ) function to populate variables.**

```
data C;
  input patno source $ gender $;
  cards;
1 C male
2 C female
```

```
3 C male
5 C male
;
run;

data D;
 input patno source $ age;
 cards;
2 D 45
2 D 40
4 D 35
;
run;

proc sql;
  create table test12 as
    select coalesce (c.patno,d.patno) as patno,     ❶
           coalesce (c.source,d.source) as source,  ❷
           c.gender,d.age  ❸
  from C as c
   full join D as d  ❹
   on c.patno=d.patno;  ❺
quit;

proc print data=test12;
run;
```

Annotated Output

```
The SAS System
        (Data C, D) (Data C) (Data D) (All data set C obs)
Obs   patno    source    gender    age

  1      1        C       male      .     (No matching D obs, age
                                           is missing)
  2      2        C       female    45    (Matching obs from C & D
                                           data sets)
  3      2        C       female    40    (Matching obs from C & D
                                           data sets)
  4      3        C       male      .     (No matching D obs, age
                                           is missing)
```

| 5 | 4 | D | | 35 | (No matching C obs, gender is missing) |
| 6 | 5 | C | male | . | (No matching D obs, age is missing) |

In Example 1.26, the result of the FULL JOIN is a match of all the matching and nonmatching observations. Since the COALESCE function is used with the *patno* variable ❶, if the value of it is missing from data set C then the SQL procedure will look at the matching observation in data set D for a nonmissing value. The AS keyword is used to create the new variables. The same will be done for the *source* variable ❷. The *gender* variable is selected from the C data set and the *age* variable is selected from the D data set because of the data set alias C and D, respectively ❸. Since this is a FULL JOIN, all of the observations from both data sets are selected ❹. The *patno* variable is used in the ON clause to match the records between the two data sets ❺.

The output is very similar to the output from the FULL JOIN in Example 1.25. The difference in the output is shown in observation 5, where there is no observation in data set C from which to get the values of *patno* and *source*. Since the COALESCE function was used, the SQL procedure went to data set D to look for the nonmissing values that are used to populate *patno* and *source* variables. Note that while this result is similar to Example 1.18, there is still a difference in the values of the common variables because of the different rules between the SQL procedure and the DATA step.

The following are good reference books and articles on the SQL procedure: *Proc SQL: Beyond the Basics Using SAS, Using the SQL Procedure in SAS Programs, Using the SQL Procedure in Data Management Applications,* and Merging Tables in DATA Step vs. PROC SQL: Convenience and Efficiency Issues.

Chapter 1. Accessing Data—Chapter Summary

INPUT Statement

Type	Example	Raw Data	
List	input name $;	tim	(basic)
Column	input name $ 1–7;	tim lou	(embedded blanks)
Formatted	input @23 dob mmddyy8.;	10/23/78	(dates, nonstandard)
Named	input @ 'used' duration;	Used 0.05	(key word)
	input car=$;	Car=Toyota	

Input of Multiple Data Rows to Single Row—Line Pointer Control (/ or #*n*)

Example	Raw Data: Multiple Rows	Data Set: Single Record		
input name $ 1–8	tim	Name	Absences	Quiz
/ absences 1–2 quiz 3–6;	1 6.0	tim	1	6.0
input #1 name $ 1–8	tim	Name	Absences	Quiz
#2 absences 1–2 quiz 3–6;	1 6.0	tim	1	6.0

Input to Multiple Data Set Records from Streaming Data—Trailing @ Control

Example	Raw Data: Streaming Data	Data Set: Structure, Multiple Records		
input name $ 1–4 @;	bill 360 1000 340 2000	Name	v2	v3
if name in ('bill');	tom 20 20 20 20	bill	360	1000
input v2 v3 @; output;		bill	340	2000
input v2 v3 @; output;				

Input to Multiple Data Set Records from Streaming Data—Trailing @@ Control

Example	Raw Data: Streaming Data	Data Set: Multiple Records	
input v1 v2 @@;	20 360 21 240 22 340	v1	v2
		20	360
		21	240
		22	340

INFILE Statement to Read Data Files

```
INFILE 'C:\MYDATA\DATA.TXT'
       DLM=' ' END=LAST FIRSTOBS=2 MISSOVER TRUNCOVER ;

INPUT NAME $8. @10 ABSENCES 2. @15 QUIZ 4.1 @21 CLASS 1.
@23 DOB MMDDYY8.;
```

Combining SAS Data Sets Using the DATA Step

DATA Step Statement: Data TEST; (statement goes here) Run;	Description
Set A; (A, n=2)	Use data set A to create data set TEST (n=2)
Set B A;(B, n=2, A, n=2)	Attach records from data set A after the records from data set B (n=4)
Set A B;(A, n=2, B, n=2) By patno;	Sort data set A and B by *patno* and save to data set TEST (n=4)
Merge C D;(C,n=4, D,n=3) /* Also Set C; Set D; */	First record from data set C is matched with the first record from data set D; process is repeated for all records (one-to-one merge) (n=4)
Merge C D;(C,n=4, D,n=3) By patno;	Records from data set C and D are matched by the *patno* variable and saved to data set TEST (match merge) (n=6[# of unique combinations of by variables])

Combining SAS Data Sets Using the SQL Procedure

PROC SQL; (statements go here) quit;	Description
	Query from single data set (similar to DATA step).
	DATA TEST6;
create table test6 as	KEEP PATNO SOURCE GENDER WEIGHT;
select patno, source,	WEIGHT = 150;
gender,	SET C;
150 as weight	WHERE GENDER='male'; RUN;
from c	PROC SORT DATA = TEST6; BY PATNO;
where gender='male'	RUN;
order by patno;	

(continued)

Combining SAS Data Sets Using the SQL Procedure (*continued*)

PROC SQL; *(statements go here)* *quit;*	*Description*
	Conventional INNER join. An inner join returns a result table for all of the rows in a table that have one or more matching rows in another table.

create table test7 as
select *
from C as c,
D as d
where c.patno=d.patno;
	/* Comparable except for common variables */
	DATA TEST7;
	  MERGE C(IN=INC) D(IN=IND);
	  BY PATNO;
	  IF INC AND IND; RUN;

	OUTER join: LEFT, RIGHT, FULL. Similar to an INNER join except that all records from at least one of the data sets are included in the new data set.

create table test9 as
select *
from C as c
left join D as d
on c.patno=d.patno;
	/* Comparable except for common variables */
	DATA TEST9;
	  MERGE C(IN=INC) D(IN=IND);
	  BY PATNO;
	  IF INC; RUN;

Comparable Terms in Both DATA Step and PROC SQL

DATA Step	*PROC SQL*	*Description*
DATA SET	CREATE TABLE	Names and defines SAS data set
VARIABLE	COLUMN	Variables names in SAS data set
RECORD	ROW	Observations within SAS data sets
FIRST DATA SET	LEFT DATA SET	First data set read
SECOND DATA SET	RIGHT DATA SET	Second data set read
MERGE	JOIN	Combine data sets
MERGE WITH BY	FULL OUTER JOIN	Combine data sets keeping all records
MERGE WITH BY AND IF (BOTH DATA SETS)	INNER JOIN	Combine data sets keeping only records from common values
KEEP	SELECT	Specifies variable names to follow
SET	FROM	Specifies name of SAS data set
BY with PROC SORT	ORDER BY	Sorts records
RUN	QUIT	End of DATA step or PROC SQL

Attributes and Data Values of Common non-BY Variables from First or Last Data Set When Combining Data Sets

Below are the assumptions required to use this table:

1. Applies when combining two or more data sets using MERGE or PROC SQL.
2. Two or more data sets have common variables.
3. Length of BY variables are the same in each data set.
4. Common non-BY variables represent common variables not used in the BY statement.
5. When using the DATA Step to combine data sets, the order of data sets is assumed to be one-to-many. Other configurations such as many-to-one or many-to-many may use the first data set value instead of the last data set value if the last data set has fewer records.
6. Although the table also applies to one-to-one merges, only the match merges are recommended.
7. PROC SQL Select statement does not select variables from separate data sets and does not use the COALESE() function.
8. Variable attributes consist of LENGTH, FORMAT, INFORMAT, and LABEL.
9. First and left represent the first data set in the merge and PROC SQL FROM statement respectively.
10. Last and right represent the last data set in the merge and PROC SQL FROM statement respectively.
11. When using PROC SQL, warning messages may appear because SAS can not keep the same variables from more than one data set.

Method	FIRST Data Set with Attribute Statements	FIRST Data Set without Attribute Statements and LAST Data Set with Attribute Statements
DATA STEP: MERGE		
Length	FIRST Data Set	FIRST Data Set
Format, Informat, Label	FIRST Data Set	LAST Data Set
Data Value	LAST Data Set	LAST Data Set
PROC SQL: SELECT		
Length	FIRST Data Set	FIRST Data Set
Format, Informat, Label	FIRST Data Set	FIRST Data Set
Data Value	FIRST Data Set	FIRST Data Set

General Key Rules:

- The LENGTH is always determined from the first/left data set.
- When using PROC SQL to combine data sets, all variable attributes (LENGTH, FORMAT, INFORMAT, LABEL) are determined from the first/left data set.
- When merging data sets, common non-BY variable values are kept from the last/right data set.
- When using PROC SQL to combine data sets, common non-WHERE variable values are kept from the first/left data set.

Chapter 1.   Accessing Data—Chapter Questions

Question 1: What effect occurs when you mix the following INFILE statement options: FIRSTOBS = 5 and OBS = 10?

Question 2: What if there is no "." to represent missing values? How can you correctly read missing values from the raw file?

Question 3: When using the data set option (IN=) in a DATA step, is the variable saved with the data set?

Question 4: Does a data set need to be presorted before it can be combined with another data set when using the SET and BY statements?

Question 5: When using the COLUMN INPUT statement, is it possible to read your data in any order?

Question 6: When merging data sets using the MERGE statement, which data set, first or second, has data values that overwrite data values for common variables?

Question 7: When using the LIST INPUT statement, is it possible to have embedded blanks in your data values?

Question 8: When setting two data sets together in a SET statement with one having five observations and another having six observations, how many observations are in the final data set?

Question 9: Using the pound-n ($\#n$) line pointer control, is it possible to skip rows when reading raw data?

Question 10: From which data set are the data values of common variables used in a new data set if merging two data sets with the MERGE and BY statements?

Question 11: When setting two data sets together with a SET statement, which data set defines the format, informat, and label for common variables?

Question 12: What informat is used to read in the date 10/23/78?

Question 13: When using the LIST INPUT statement, is it possible for the data values to be separated by more than one blank or delimiter?

Question 14: For a given SQL procedure code, write the same corresponding process using the DATA step except for common variables.

```
proc sql;
  create table test7 as
  select *
  from C as c,
  D as d
  where c.patno=d.patno;
  quit;
```

Question 15: Why is the OUTPUT statement generally applied when using multiple INPUT statements?

Question 16: When using the BY statement in a DATA step, are the FIRST.by_variable and LAST.by_variables saved with the data set?

Question 17: What is the difference between LEFT, RIGHT, and FULL joins and how is the OUTER join different from the INNER join?

Question 18: What is the equivalent DATA step for performing a OUTER FULL join?

Question 19: When using the INFILE statement, what is one way to read commas as part of the data value?

Question 20: When using the SET or MERGE statement with the BY statement, should the data set be presorted?

Question 21: From which data set are the attributes of common variables used to specify the new data set if merging two data sets with the MERGE and BY statements?

Question 22: Which clause in the SQL procedure is used to select variables?

Question 23: When using a FORMATTED INPUT statement, can a variable without an informat be specified?

Question 24: When reading data, if you do not specify the length of variables, what is the default length?

Question 25: Can the trailing @ control be used in the LIST, COLUMN, or FORMATTED INPUT statements?

Question 26: When merging two data sets with the MERGE and BY statements, what is the total number of records in the final data set?

a) Sum of all data sets
b) Data set with minimum number of observations
c) Data set with maximum number of observations
d) Data set with number of unique combinations of the BY variables

Question 27: By default, how are the missing numeric and missing character data represented in SAS data sets?

Question 28: Name at least two things the SQL procedure can be used for.

Question 29: What is the equivalent DATA step for performing a conventional INNER join except for common variables?

Question 30: Can the double trailing @@ control be used in the LIST, COLUMN, or FORMATTED INPUT statement?

Question 31: Are character data case-sensitive? For example, is "Sally" equivalent to "sally"?

Question 32: How do you handle multiple delimiters in the raw file when reading the raw data file?

Question 33: When can you expect to get similar results if using the MERGE statement and multiple SET statements to merge data sets?

Question 34: Name two reasons for using the INFILE statement option DSD with a LIST INPUT statement?

Question 35: When using the SQL procedure to combine tables, what is the maximum number of tables that can be combined at once?

Question 36: Is it possible to specify the length of character or numeric variables in any of the three types of INPUT statements (LIST, COLUMN, FORMATTED)?

Question 37: In which method, MERGE with BY statement or SQL procedure, will you get a WARNING message when combining common variables from data sets unless you select variables from individual data sets?

Chapter 2

Creating Data Structures

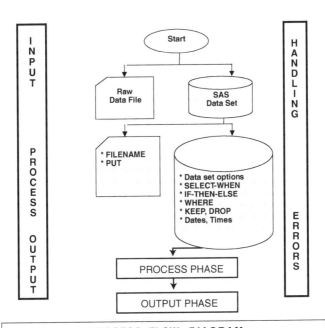

2.1   Introduction

In this chapter, you will learn how to create temporary and permanent SAS data sets. Temporary data sets are deleted at the end of a SAS job or session, while permanent data sets are saved for future use. After covering how data sets are created, you will discover how to control which observations you want to keep in the new data set and how to assign values conditionally to new and old variables. You will also learn how to export (or write) your data set to an external file. The last section of this chapter discusses how the SAS DATA step is compiled and executed.

2.2   Creating Temporary and Permanent SAS Data Sets

2.2.1   General Rules

One of the benefits of creating temporary and permanent SAS data sets from another data set is to prevent changing or damaging the original data set. The DATA statement is used to assign the name of the SAS data set. Another function of the DATA statement is to create the data set as either a temporary or a permanent data set depending how the data set name is specified.

When the data set is created, there are two major components that make up the data set. The first component is the descriptor portion, which contains the information about the data set contents. The second component of the data set is the data portion, which contains the actual values as observations (rows) and variables (columns). Note that SAS enables you to access the power of the DATA step without creating an SAS data set by specifying the keyword _NULL_. See Section 2.4 for more information.

2.2.2   Creating Permanent Data Sets

Syntax:
   LIBNAME <library reference> 'physical location';

   DATA <library reference>.<data set name>;
      SET <data set name>;
   RUN;

Using the DATA statement followed by the two-level name creates a permanent data set. The first part of the name will specify the libref or library reference. The libref is a shorthand name used to reference the location where the data set will be saved. The libref is created with a LIBNAME statement. A LIBNAME statement associates the libref name with the data library where the data set will be saved. Once the LIBNAME statement is specified, any SAS procedure or DATA step can reference the libref and data set name. **LIBNAME statements are global and remain in effect until you change them or cancel them.** A data library can contain multiple SAS data sets along with other SAS files. A data set created with a libref is always available and can be accessed by referencing the libref. Using a two-level name does not always guarantee a permanent data set is created. You can create a temporary data set by specifying a two level name, using WORK as the libref. As mentioned in Example 1.8, if the OUTPUT statement is not specified, then an implied OUTPUT statement is included at the end of the DATA step.

Example 2.1   Creating a permanent data set.
```
libname mylib 'c:\mydata'; ❶

data mylib.center mylib.phone; ❷
    set dsname;
 /* Implied Output statement */
run;
```

In Example 2.1, the libref mylib ❶ is created with the LIBNAME statement and specifies where the data set will be saved; in this case, it will be the mydata directory on the C drive. The DATA statement will specify the two-level name for the data set. In this case, the two data set names are specified mylib.center and mylib.phone ❷. Since two names are specified, two exact permanent data sets will be created from the DSNAME data set.

2.2.3   *Creating Temporary Data Sets*

Syntax:
> **DATA <data set name>;**
> > **SET <data set name>;**
> **RUN;**

The temporary data set is also created with the DATA statement when only the name of the data set follows the DATA statement. When creating temporary data sets, the libref is implied. **The WORK libref is the default implied library for temporary data sets** and does not need to be specified in the DATA statement. Temporary data sets are available only for the duration of the SAS session and must be recreated with each run. If you do not specify a data set name, then SAS creates a default name of DATAn, where n is a consecutive numbering of the data sets. The first data set created without a name would be called DATA1 and the next DATA2 and so on. The temporary data sets or data sets in the WORK library will be deleted automatically when the SAS session is terminated.

Example 2.2 Creating a temporary data set.

```
data demog;
        set dsname;
    /* Implied Output statement */
run;
```

In Example 2.2, the temporary data set does not need a LIBNAME statement since the WORK directory is the default implied library. Also, since this is a temporary data set, only the data set name needs to follow the DATA statement. The demog data set is created from the DSNAME data set.

2.2.4 Using Data Set Options

Syntax:

DATA <data set name> <(data set option)>;

SET <data set name> <(data set option)>;
RUN;

or

SAS Procedure DATA = <data set name> <(data set option)>;
RUN;

Now that you know how to create data sets, you should know that there are many data set options available for processing the data set. Data set options are placed in parentheses just after the data set name. Data set options placed in statements such as the SET statement occur when the records are being read while data set options placed in statements such as the DATA statement occur when the records are being written. **In addition, data set options can be specified in SAS procedures.** Some of the options listed in Table 2.1 have already been introduced. Many of these data set options can also be used as SAS statements in the data step. **Note that one or more options may be specified for combined effects and that options can be listed in any order.** When specifying multiple data set options, separate options with at least one space and enclose all options separated by spaces in a single set of parentheses.

Example 2.3 Using data set options.

```
proc print data =test (obs=20);
run;
```

Table 2.1 Data Set Options

Data Set Option	Example	Description
(DROP=)	(DROP = A B)	Removes variables A and B from data set. Use the original variable name if used with the RENAME data set option.
(FIRSTOBS=)	(FIRSTOBS = 5)	Starts reading from the specified observation number (5).
(IN=)	(IN = DS1)	Creates temporary variable DS1 to identify source data set.
(LABEL = ())	(LABEL = (A = 'Age'))	Relabels variable A to Age.
(KEEP=)	(KEEP = A B)	Saves variables A and B to data set. Use the original variable name if used with the RENAME data set option.
(OBS=)	(OBS = 10)	Stops reading after the specified observation number (10).
(RENAME = ())	(RENAME = (A = D))	Changes variable name from A to D. Note that other data set options should reference the original variable name and not the new name.
(WHERE = ())	(WHERE = (A >= 20))	Subsets data set to include all A >= 20. Note that variables specified in the WHERE condition should not be included in the DROP= data set option.

In Example 2.3, SAS displays the first 20 records from the test data set because of the OBS= data set option. Using data set options with SAS procedures offers great flexibility without having to first preprocess the data set with a DATA step.

2.3   Applying Conditional Assignments and Conditional Subsets

Conditional assignments are statements in a DATA step that will create a variable or modify a variable based on the values of another variable or the same variable. The conditional statements can be applied to both character and numeric variables and are executed when saved to the final data set. When using conditional assignments, make sure that the expression is appropriate for the variable type to prevent SAS from generating an ERROR message.

In addition, remember that **character data is case sensitive and may contain spaces, numbers, or symbols.**

Conditional subsets are used to reduce the number of observations in a data set based on selected values of variables. Instead of another variable being assigned values, SAS uses these statements to exclude observations that do not meet the specified condition of the variable. As an alternative, you can exclude observations that meet the specified condition when applying the DELETE statement.

2.3.1 Conditional Assignments

Syntax:

```
/*  Select-When Conditional Assignment */
  SELECT <(select expression)>;
    WHEN <(when expression 1)>;
    ...
    WHEN <(when expression n)>;
    OTHERWISE <otherwise statement>;
  END;

/* If-Then Conditional Assignment */
 IF <expression> THEN <statement>;
 <ELSE IF <statement>>;
 <ELSE <statement>>;
```

The first type of conditional assignment is the SELECT–WHEN statement. The SELECT–WHEN statement defines **mutually exclusive conditions to assign a second variable.** When the value of the first variable matches the value specified, the value for the second variable is assigned. Any valid expression will work as long as it is appropriate for the value type. Note that expressions may include arrays or functions. Although it is not required, the OTHERWISE clause is recommended to assign values for unexpected variable values.

The second type of conditional assignment is the IF–THEN statement. The IF–THEN statement assigns the value of the second variable based on the value of the first variable. With the addition of the ELSE clause, the **IF–THEN–ELSE statement becomes a mutually exclusive statement.** As with the SELECT–WHEN statement, any valid expression will work as long as it is appropriate for the value type.

If you are specifying only a few conditional assignments, then the IF–THEN–ELSE statement should be used. If there are many conditional assignments, then the SELECT–WHEN statement should be used since it is more efficient and will thus reduce the CPU time. The SELECT–WHEN statement is also easier to read when there are many conditional assignments. Remember that if any of the required clauses are missing from the SELECT–WHEN or IF–THEN–ELSE statements, SAS will issue a syntax ERROR message.

Example 2.4   Applying conditional assignments.

```
data test13;
   input name $ class $ test_score ;
   cards;
   Tim math 9
   Tim history 8
   Tim science 7
   Sally math 10
   Sally science 7
   Sally history 10
   John math 8
   John history 8
   John   science 9
;
run;

data grade;
   set test13;
   if class='math' then classnum=1;
   else if class='science' then classnum=2; ❶
   else if class='history' then classnum=3;
select(test_score);
      when(10) grade='A';
      when(9) grade='B';   ❷
      when (8) grade='C';
      otherwise grade='D';
   end;
run;

proc print data=grade;
run;
```

Output

Obs	name	class	test_ score	classnum	grade
1	Tim	math	9	1	B
2	Tim	history	8	3	C
3	Tim	science	7	2	D
4	Sally	math	10	1	A
5	Sally	science	7	2	D

6	Sally	history	10	3	A
7	John	math	8	1	C
8	John	history	8	3	C
9	John	science	9	2	B

In Example 2.4, the IF–THEN–ELSE statement is used to assign a value for the *classnum* variable based on the *class* variable ❶. The SELECT–WHEN statement is used to assign a value for the grade variable based on the value of the *test_score* variable ❷. Note that because these are mutually exclusive statements, SAS saves time by not checking the remaining values once a match is found. Note that for values of the *test_score* variable other than 10, 9, or 8, the grade variable gets assigned to D.

2.3.2 Selecting Observations with the WHERE/IF Statements

Syntax:
DATA Step Statement:
   WHERE <where expression>;
   IF <if expression>;
RUN;
Data Step Option:
   (WHERE=(<where expression>));
   If statement is not valid
RUN;

PROC Statement:
      As a statement: WHERE <where expression>;
            Or
      As a data set Option: (WHERE=(<where expression>));
RUN;

The WHERE and IF keywords can both be used for subsetting data sets. There are some differences between the WHERE and IF keywords regarding when they can be used and which options can be used with them. Both the IF and WHERE statements can be used in the DATA step to subset your data. As seen in Figure 2.1, **WHERE conditions are applied BEFORE the data enters the input buffer while IF conditions are applied after the data enters the PDV. In addition, both IF and WHERE statements are cumulative statements in that additional IF or WHERE statements can be added to the subset condition.** For large data sets, you will notice that the WHERE statement is more efficient since the data is subsetted before being read into the program data vector. There are two

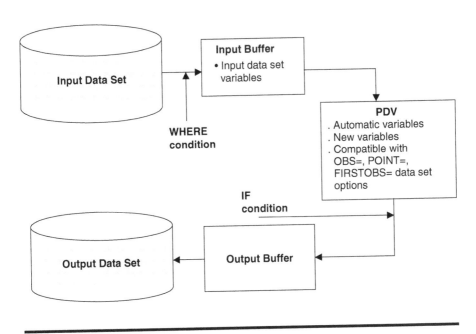

Figure 2.1   WHERE and IF conditions applied.

cases where the WHERE statement cannot be used in a DATA
step. **The WHERE statement can only be used with variables
that exist in the input data set.** If the WHERE expression
contains a variable that is created in the DATA step or a temporary
variable such as _N_, SAS will generate an ERROR message. In
addition, the WHERE statement cannot be used with the INPUT
statement or these data set options: POINT= and FIRSTOBS=. A
SET, MERGE or UPDATE statement must exist in the DATA step
in order to use the WHERE statement. **Note also that when
merging data sets, the WHERE condition is applied before
merging data sets as opposed to applying after merging the
data sets when specifying the IF condition. Depending on
the data sets being merged, this difference could cause inconsistent results.**

Another major difference between the WHERE and IF conditions is
that only the WHERE keyword can be used as a data set option or in
SAS procedures. The syntax for the WHERE option is a little more complex
than the WHERE statement since it requires the correct placement of the
parentheses and the equal sign. There are no efficiency differences
between the WHERE option and WHERE statement. The WHERE statement
and the WHERE option both can be used with SAS procedures. Note that

Table 2.2   WHERE Operators

Operator	Description
BETWEEN–AND	Includes values defined in the range or numeric variables; e.g., WHERE AGE BETWEEN 18 AND 34; this is similar to AGE >= 18 AND AGE <= 34 or 18 <= AGE <=34
COLON MODIFIER (:)	Compares shorter text with longer text by truncating the longer text to the length of the shorter text; default is to pad shorter text with trailing blanks in order to make the comparison; e.g., WHERE NAME = : 's';
CONTAINS or ?	Used to search for a specific text in character variables, e.g.,WHERE NAME CONTAINS 'ally' or WHERE NAME ? 'ally';
IS NULL or IS MISSING	Includes all missing values including special missing values. Note that the variable can be character or numeric. e.g., WHERE NAME IS MISSING;
LIKE 'PATTERN'	In character variables, used to search for similar match. Use with the underscore (_) or the percent sign (%) operator; e.g., WHERE NAME LIKE 's%';
PERCENT SIGN (%)	Any number of characters are possible, similar to a wildcard character; e.g., WHERE NAME = 's%'; WHERE NAME ? 'ally';
SOUNDS-LIKE (SOUNDEX)	"=*" includes all similar character values that sound alike. This does not require the exact spelling of the character value; e.g., WHERE NAME =* 'sal';
UNDERSCORE (_)	Each underscore represents any single character; e.g., WHERE NAME = 's____';

if the WHERE option is specified on the new data set along with a WHERE statement within the DATA step, then SAS ignores the WHERE statement and applies the WHERE data set option.

There are many operators in SAS that can be used with both the IF and WHERE keywords. There are some operators that can only be used with the WHERE keyword. Table 2.2 contains a list of operators that can be used only with the WHERE keyword.

Starting with SAS Version 8.1, the WHERE or IF statements can be used with the OBS= data set option. The OBS= data set option is used to select a number of observations that are specified in the option. If the OBS= data set option is used with the WHERE or IF statement, it is important to note that there will be different results depending on which statement is specified. Due to this, it is not recommended to use the OBS= data set option with the WHERE or IF statement. In addition, in general it is not recommended to specify both WHERE and IF conditions in the same DATA step to subset the data set. Finally, as seen in Example 2.5, the IF statement can be used to assign new variables while the WHERE statement can only subset the data set. See the SAS paper *How and when to Use WHERE* for information about the WHERE statement.

Example 2.5 Applying conditional subsets.

```
data test13;
   input name $ class $ test_score ;
   cards;
   Tim math 9
   Tim history 8
   Tim science 7
   Sally math 10
   Sally science 7
   Sally history 10
   John math 8
   John history 8
   John science 9
;
run;

data grade;
   set test13;

   where name='Tim' or name='Sally'; ❶

   if class='math' then classnum=1;
   else if class='science' then classnum=2; ❷
   else if class='history' then classnum=3;

   if classnum=2; ❸
  run;

proc print data=grade;
run;
```

Output

Obs	name	class	test_ score	classnum
1	Tim	science	7	2
2	Sally	science	7	2

Example 2.5 shows the use of both the IF and WHERE statements to select observations. The WHERE statement subsets the data based on the value of the *name* variable ❶. The WHERE statement was used since it is more efficient. The next set of IF statements create the *classnum* variable ❷. Because the *classnum* variable is created in the data set, the IF statement is required when specifying the same variable to subset the data set ❸. The WHERE statement cannot be used on variables that do not exist in the input data set. Note that the final data set contains only the records that meet both subset conditions on the *name* and *classnum* variables.

2.3.3 *Variable Processing*

Syntax:

DATA step statements — occurs when writing or processing output data set:
  DROP <variable 1...variable n>;
  KEEP <variable 1... variable n>;

Input data set options — occurs when reading input data set
Output data set options — occurs when writing to output data set
  (DROP = <variable 1... variable n>)
  (KEEP = <variable 1 ... variable n>)

The DROP and KEEP statements are used to specify which variables are saved or not saved in the final data set. Both statements can be used as a DATA step statement or DATA step option. If they are used as a DATA step statement, then the selection occurs at the writing or processing of the output data set. If they are used as data set options for the input data set, then the selection occurs at the reading of the input data set. If they are used as data set options for the output data set, then the selection occurs at the writing of the output data set. The best guidance for using the DROP or the KEEP statement is to apply the statement that requires the shortest list of specified variables. Generally, it is best not to use both statements within the same DATA step. **If both statements are applied together, the order of priority is DROP and KEEP respectively.** To help create variable lists to be used in the DROP and KEEP statements,

Table 2.3 **Variable List Shortcuts**

Shortcut	Expanded Equivalent	Description
DROP A - - D;	DROP A B C D;	Lists variables based on the order stored in the data set from variable A to variable D. This is called named range.
KEEP _ALL_;	KEEP all variables;	Lists all variables in the data set.
KEEP _CHAR_;	KEEP all character;	Lists all character variables in the data set.
KEEP _NUMERIC_;	KEEP all numeric;	Lists all numeric variables in the data set.
VAR ROOT1 – ROOT3;	VAR ROOT1 ROOT2 ROOT3;	Lists variables from starting number to ending number based on root name.

SAS offers the following shortcuts in Table 2.3. Note that variable list shortcuts can also be specified in VAR statements. See Chapter 3 for more information on using the VAR statement in SAS procedures.

Example 2.6 **Using the DROP statement to process variables.**

```
data test13;
   input name $ class $ test_score ;
   cards;
   Tim math 9
   Tim history 8
   Tim science 7
   Sally math 10
   Sally science 7
   Sally history 10
   John math 8
   John history 8
   John science 9
;
run;

data grade;
   set test13;
   drop class;
run;
```

```
proc print data=grade;
run;
```

Output

Obs	name	test_ score
1	Tim	9
2	Tim	8
3	Tim	7
4	Sally	10
5	Sally	7
6	Sally	10
7	John	8
8	John	8
9	John	9

In Example 2.6, the DROP statement is used instead of the KEEP statement, since the number of variables dropped is shorter then the number of variables kept. As an alternative, the DROP option could have been applied to the data set TEST13 or grade to give the same result.

2.3.4 Creating and Manipulating SAS Date/Time Values

SAS saves dates as integer values in numeric variables. The values are calculated as the **number of days from January 1, 1960 to the given date.** January 1, 1960 has an SAS date value of 0. Figure 2.2 shows how calender dates are assigned to SAS date values. In general, an INFORMAT is used to read the date and a FORMAT is used to display the date, e.g., the DDMMYY. or DDMMYYYY. When creating a date variable and setting it to a constant value, the value must be assigned in the format 'DDMMMYY' or 'DDMMMYYYY,' where the text is enclosed in single or double quotes and followed with the letter D. Because dates are stored as numbers, any mathematical function can be performed on date variables.

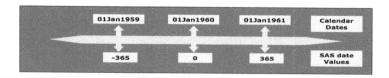

Figure 2.2   SAS dates.

Time variables are stored in SAS in the same manner as date values, except that time values are always positive. **An SAS time is stored as the number of seconds from midnight. As with date variables, an INFORMAT is used to read the time and a FORMAT is used to display the time.** In order to create a time variable with a constant value, the value must be assigned in the format 'HH:MM:SS' or 'HH:MM,' where the text is enclosed in a single quote and followed with the letter T.

Note that there also is a type of variable that combines both the date and the time, called a *datetime* variable. A SAS *datetime* value is stored as the number of seconds between midnight on January 1, 1960 and a given date and time. In order to create a *datetime* variable with a constant value, the value must be assigned in the format 'DDMMMYY:HH:MM:SS' where the text is enclosed in a single quote and followed with the letters DT. The year in the format can be a two or four digit year and the time can be with or without the seconds.

If two-digit years are specified in the code or exist in the raw data file, then the SAS system option YEARCUTOFF= determines what the first two digits will be. The default value for the YEARCUTOFF= option is 1920. The year 1920 specifies the first year of the 100-year span used to determine the first two digits. **So, in effect, with the default YEARCUTOFF= option, the 100-year span would be 1920 to 2019.** Therefore, any dates with two-digit years between 20 and 99 will use 19 as the prefix and any dates with two-digit years between 00 and 19 will use 20 as the prefix, unless the YEARCUTOFF value is changed. All four-digit years are always read correctly. See the SAS paper *Blind Dates and Other Lapses of Reason: Handling Dates in SAS* for more information on SAS dates.

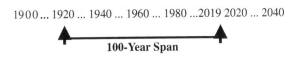

1900... 1920 ... 1940 ... 1960 ... 1980 ...2019 2020 ... 2040

100-Year Span

Example 2.7   Creating and manipulating SAS dates.

```
data class3;
   input name $8. @10 absences 2. @15 quiz 4.1 @21
class 1. @23 dob mmddyy8.;
cards;
tim lou  1    6.0    3 10/23/78
sally    2   10.1    2 01/02/78
john     2    8.0    3 06/22/78
;
run;
```

```
data dates;
  length nxyrdt 4.;
  set class3;
  newdt = '01jan2002'd;      ❶    /* Assign a date constant */
  nxyrdt = newdt + 365;      ❷    /* Date calculation */
  if dob gt '02feb1978'd;    ❸    /* Compare dates */
  format dob newdt nxyrdt mmddyy8.; /* Format dates */
run;

proc print data=dates;
run;
```

Output

Obs	nxyrdt	name	absences	quiz	class	dob	newdt
1	01/01/03	tim lou	1	6	3	10/23/78	01/01/02
2	01/01/03	john	2	8	3	06/22/78	01/01/02

Example 2.7 shows how SAS dates and times can be created and manipulated. The *newdt* variable is created as a date constant where the date is specified in the 'DDMMMYYYY' format ❶. Since dates are stored internally as the number of days from January 1, 1960, you can perform any mathematical function on them. The *nxyrdt* variable is created by adding 365 days to the value of the *newdt* variable, giving a new date of 01/01/03 ❷. In the last part of the example, the *dob* variable is compared to a constant date ❸. Only the two records that have *dob* on or after 02 FEB 1978 are selected. In order to understand the values of the date variables, all of the date variables have been assigned a format of mmddyy8.

2.4 Using DATA Step Statements to Export (or Write) Data to a File

SAS can import and export data from external files. Chapter 1 shows how to create data sets from external files. This section shows how to export data from SAS to an external file.

2.4.1 Using DATA Step Statements to Export Data to a File

Syntax:

FILENAME <file reference> '<physical file location>' <options>;
DATA _NULL_;
  FILE <file reference>;

```
  SET data set;
  PUT <>;
RUN;
```

The FILENAME statement is used to assign a libref or library reference to the external file. FILENAME statements are global and remain in effect until you change them or cancel them. The reference can contain the full path of the file along with the file name. **The word _NULL_ that appears in the DATA statement is a special keyword that instructs SAS not to create a data set.** The FILE statement uses the file reference that was created with the FILENAME statement. The FILE statement is analogous to the INFILE statement that was used to read external data. Some of the INFILE statement options such as DLM= and DSD are also available on the FILENAME statement. Use the DLM= option to create delimited text files and the DSD option to have commas enclosed in quotation marks. The SET statement instructs SAS which data set is used when creating the external file.

The PUT statement specifies the variables to write, the formats to apply, and the order in which the variables are written to the external file. The PUT statement is analogous to the INPUT statement that is used to read external data. If a PUT statement is used without a FILE statement, the variables will be written to the SAS log file. This is one effective technique to use the PUT statement to display data for debugging purpose. For example, using the PUT _N_ statement in an IF–THEN statement will display the record number when the specified condition is met. See Example 5.12 for more information on debugging techniques using the PUT statement.

When writing to the file, SAS offers several options in the PUT statement as seen in Table 2.4. The _ALL_ keyword displays all variables as NAME = value, while the _INFILE_ keyword just displays all values.

Table 2.4 PUT Statement Options

PUT Term Type	Example
Character constant	PUT 'My comment ' 'plus another comment';
Column control	PUT NAME $8–15;
Format	PUT DOB DATE9.;
Keyword	PUT _N_; PUT _ALL_; PUT _INFILE_;
Pointer control	PUT @10 'My comment';
Variable term	PUT @10 my_comment=;

Example 2.8   Writing data to a file from a data set.

```
data class3;
   input name $8. @10 absences 2. @15 quiz 4.1 @21 class
   1. @23 dob mmddyy8.;
   cards;
```

```
tim lou   1     6.0    3 10/23/78
sally     2    10.1    2 01/02/78
john      2     8.0    3 06/22/78
;
run;


FILENAME outfile 'c:\OUTPUT_FILE.TXT';
DATA_NULL_;
   FILE outfile; ❶
   SET class3;
   PUT 'NAME = ' NAME $ 8-15 ' BIRTH DATE = ' DOB DATE9.
   +2 QUIZ 4.1 +2 CLASS=; ❷
RUN;
```

Output — 'OUTPUT_FILE.TXT'   ❸

```
NAME = tim lou   BIRTH DATE = 23OCT1978    6.0   CLASS=3
NAME = sally     BIRTH DATE = 02JAN1978   10.1   CLASS=2
NAME = john      BIRTH DATE = 22JUN1978    8.0   CLASS=3
```

Example 2.8 creates the external file OUTPUT_FILE.TXT that is specified in the FILENAME statement using the file reference OUTFILE. The same file reference, OUTFILE, is used in the FILE statement in the DATA step. ❶ The PUT statement specifies which variables are exported along with the formats and order of the variables. The PUT statement also specifies the variable class with an equal sign. ❷ This instructs SAS to display the variable name, the equals sign, and the variable's value. ❸

2.5 Exporting Data to Excel and Access Using the EXPORT Procedure

With SAS, it is easy to export data to Excel or Access using the EXPORT procedure. The approach shown in this section creates a comma delimited text file created from SAS. The EXPORT wizard can be used to automate this process and automatically create the SAS code.

To export data directly, without the comma delimited text file, to Excel or Access files requires the SAS/ACCESS to PC module. The only change required would be to specify the Excel file name in the OUTFILE= option. For directly exporting data to Access tables, without the tab delimited text file, requires the Access table name in the OUTTABLE= option. An alternative to using the EXPORT procedure to create an Excel file is to create the Excel file from ODS. See Example 4.10 for more information.

Syntax
PROC EXPORT DATA= <data set name>
        OUTFILE= "<physical file location>"
        DBMS=<CSV I TAB> <REPLACE>;
RUN;

Example 2.9   Create an Excel file from a comma separated file.
```
proc print data=class;
run;
```

```
proc export data= work.class
        outfile= "c:\mydata\sas_export_class_excel_file.csv" ❶

    dbms=csv replace;

run;
```

Output
```
The SAS System
```

Obs	name	absences	quiz	class	weight	❷
1	tim	1	6	3	.	
2	sally	2	10	4	110	
3	john	-2	8	3	120	

```
CVS file – SAS_export_class_excel_file.csv ❸

name,absences,quiz,class,weight  ❹
tim,1,6,3,
sally,2,10,4,110
john,-2,8,3,120
```

The first step in Example 2.9 is to export the data set as a comma separated file using the EXPORT procedure ❶. The EXPORT procedure will create the comma separated file, sas_export_class_excel_file.csv, from the class data set, since the DBMS= CSV option is specified. The data set has three records and five variables ❷. The REPLACE option instructs SAS to replace the file if it already exists ❸. You can use the EXPORT wizard to automatically create this code. The variable names are in the first row ❹. Excel will automatically read and convert this comma separated file to an Excel file. The Excel file contains all values in the data set.

As an alternative to the EXPORT procedure, the FILE and PUT state-ments could also create comma separated files. See Example 2.8 for more information. In addition, another method to export data to Excel files is the ODS approach. See Example 4.10 to create the HTML file. To create an Excel file, all that is required is to use the HTML destination with .xls as the filename extension.

Example 2.10   Create an Access table from a tab delimited file.

```
proc print data=class2;
run;

proc export data= work.class2
            outfile= "c:\mydata\sas_export_class.txt" ❶
            dbms=tab replace;
run;
```

Output

```
The SAS System
```

Obs	name	absences	quiz	class	weight	❷
1	tim	1	6	3	.	
2	sally	2	10	4	110	
3	john	-2	8	3	120	

```
Tab file — SAS_export_class.txt ❸
```

name	absences	quiz	class	weight	❹
tim	1	6	3		
sally	2	10	4	110	
john	-2	8	3	120	

The first step in Example 2.10 is to export the data set as a comma separated or tab delimited file, such as sas_export_class.txt, using the EXPORT procedure ❶. The data set has three records and five variables ❷. The EXPORT procedure will create the tab delimited file, sas_export_class.txt, from the class2 data set since the DBMS= TAB is specified. The REPLACE option instructs SAS to replace the file if it already exists ❸. You can use the EXPORT wizard to create this code automatically. The variable names are in the first row ❹. Access can read and import the comma or tab delimited text file to create an Access table. You will need to answer questions about the structure of the text file. Make sure to select first row as variable names. The Access table will contain all values in the data set. As an alternative to the EXPORT procedure, the FILE and PUT statements could also create tab delimited files. See Example 2.8 for more information.

2.6 Understanding How the DATA Step Is Compiled and Executed

The DATA step is one of the most powerful tools in the SAS system. The two phases of the DATA step processing are the compilation phase and the execution phase. Understanding the two phases of the DATA step will facilitate debugging your program and will enable you to have confidence in your results. Table 2.5 summarizes the compile and execute steps that will be reviewed in detail in this section.

2.6.1 Compilation Phase

Example 2.11 and Example 2.12 will be used to illustrate the compilation and execution phases of the DATA step. Example 2.11 creates a data set from a raw data file, while Example 2.12 creates a data set from another data set.

Table 2.5   Summary of Compile and Execute Phases

Compile Steps	Description
1	Compile all SAS statements
2	Check syntax of all statements in the DATA step
3	Create an input buffer
4	Create a program data vector (PDV)
5	Create the descriptor portion of the new SAS data set

Execute Steps	Description
1	Read the first record into the input buffer
2	Initialize the PDV
3	For each record, execute the DATA step statements sequentially
4	Write the values from the PDV to the output data set
5	Loop back to the top of the input data set
6	Read in the next sequential observation
7	Continue to execute steps 1–6 until the end-of-file marker is reached
8	Close the output data set

Example 2.11   Create data set from reading data from a raw file.

```
c:\mydata\data.txt
12345678901234567890123456789 0
tim lou   1     6.0   3 10/23/78
sally     2    10.1   2 01/02/78
--------------------------------------------------------
data dsname;

  infile 'c:\mydata\data.txt' dlm= ' '   end=last firstobs=2
         missover truncover;
  input name $8. @10 absences 2. @15 quiz 4.1 @21 class
  1. @23 dob mmddyy8.;

/* PDV - SAS guts */
if absences = 1 then absences = 0;
/* output statement (default) */
run;
```

The compilation phase is the first time that SAS will pass through your code. If SAS encounters an error, then the data set will not be created. **The compilation phase consists of five steps:**

1. **Compile all SAS statements into machine code.** When SAS compiles the SAS statements, it is translating the statements into machine code.

2. **Check syntax of all statements in the DATA step.** SAS searches for keywords and flags only text not recognized.

3. **Create an input buffer.** The input buffer is the area in memory that will be the raw data storage (holds a record from the external file if using INFILE and INPUT statements). No data values are processed at this time.

4. **Create a program data vector (PDV).** The PDV is a space in memory where values and attributes for each variable in the output record are temporarily stored during the execution of the DATA step. All variables are placed into the PDV unless the DROP or KEEP statements are used as DATA step statements or data set options to select variables. If the RENAME= data set option is used on the input data set, then the new variable name is used in the PDV and the resulting data set. If the RENAME= data set option is used on the resulting data set, the original variable name is used in the PDV and the new variable name is used in the resulting data set. Although arrays have not yet been discussed, if arrays are used, then the variables specified in the array statement are created. Also, any new calculated variables will be created at this time as determined from the reference of the

first variable in the DATA step. In addition to the data set variables, two automatic temporary variables, _N_ and _ERROR_, are created which provide information about the data processed. Additional temporary variables may be created due to the use of a BY statement (FIRST.by_variable and LAST.by_variable) or the IN= data set option.

5. **Create the descriptor portion of the new SAS data set.** The descriptor portion of the SAS data set contains information from the PDV such as the data set attributes and the variable attributes. All variables from the PDV except the temporary variables, such as _N_ and _ERROR_, are saved in the data set. Note that at this point, data values have not yet been processed.

In Example 2.11, during the compilation phase, SAS will first read in the SAS statements and check the syntax of the statements. SAS will then create the input buffer.

INPUT BUFFER

1	2	3	4	5	6	7	8	9	0	1	2	3	4	5	6	7	8	9

After the input buffer is created, SAS will create the PDV. The PDV will list all of the variables and any temporary variables. Note that in this example, the variable attributes in the PDV are specified in the INPUT statement. In addition, all numeric variables have the default length of 8 bytes. There are five variables listed in the INPUT statement, which will be created in the same order in the PDV. If other compile-time statements, such as ATTRIB, FORMAT, INFORMAT, or LENGTH, are specified before the INPUT statement, then SAS uses these statements to define variable attributes in the PDV. Note that for character variables, SAS uses the first LENGTH or ATTRIB statement when multiple LENGTH or ATTRIB statements are specified. For numeric variables, however, SAS uses the last LENGTH or ATTRIB statement when multiple LENGTH or ATTRIB statements are specified. See the chapter summary section for a list of compile time statements. Some statements can be specified anywhere in the DATA step while others cannot if that statement must proceed other SAS statements. For example, the ARRAY statement must proceed any other SAS statement that references the ARRAY elements. In general, it is not recommended to specify multiple LENGTH or ATTRIB statements for each variable.

PROGRAM DATA VECTOR

NAME	NAME	ABSENCES	QUIZ	CLASS	DOB	_N_	_ERROR_
TYPE	$	N	N	N	N	N	N
LENGTH	8	8	8	8	8	8	8
VALUE							

The descriptor information about the data set and the variables will be created at this time. The information about the temporary variables is not saved.

DSNAME: Descriptor Information

NAME	NAME	ABSENCES	QUIZ	CLASS	DOB
TYPE	$	N	N	N	N
LENGTH	8	8	8	8	8
VALUE					

2.6.2 Execution Phase

The execution phase is the second pass through the code, the pass in which the SAS statements will be executed. **In general, each SAS DATA step statement will execute once for each observation that is read.** After successfully completing the compilation phase, SAS creates a SAS data set. If there are execution errors or run-time errors, then the data set may have incomplete data.

The execution phase consists of the following eight steps.

1. The first record is read from the external file or the input data set to the input buffer.

INPUT BUFFER (Partial)

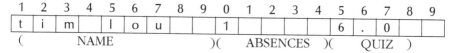

1	2	3	4	5	6	7	8	9	0	1	2	3	4	5	6	7	8	9
t	i	m		l	o	u			1					6	.	0		

(NAME   )(ABSENCES   )(QUIZ   )

2. Initialize the PDV. All the variables in the PDV are initialized to missing **before** each execution of the DATA step. If a RETAIN statement is used, the values for the variables specified in the RETAIN statement are preserved. In addition, if the COUNT+1 statement is used, the value of *count* is preserved. The values of the variables are read into the PDV sequentially from the input data set or the input file. SAS will populate each of the temporary variables, such as _N_ and _ERROR_, based on the data processed. For example, the value in the _N_ variable increments with each new observation read. The _ERROR_ variable is set to 1 for a data error in that observation. If an IN= data set option is used, this variable will be set to 1 when the current observation is read and to 0 when the current observation is not read from the data set.

PROGRAM DATA VECTOR

NAME	NAME	ABSENCES	QUIZ	CLASS	DOB	_N_	_ERROR_
TYPE	$	N	N	N	N	N	N
LENGTH	8	8	8	8	8	8	8
VALUE			.		.		.	.	.	.

3. For each record, execute each DATA step statement sequentially. The values of the input buffer are first written into the PDV.

PROGRAM DATA VECTOR

NAME	NAME	ABSENCES	QUIZ	CLASS	DOB	_N_	_ERROR_
TYPE	$	N	N	N	N	N	N
LENGTH	8	8	8	8	8	8	8
VALUE	tim lou	1	6.0	3	6870	1	.

Although the dob variable is a date variable, it is stored in the PDV as a number. After the values are written into the PDV from the input buffer, the executable SAS statements are read. The values stored in the PDV can be changed during the DATA step execution process when specifying conditional assignment statements such as the IF–THEN statement.

In Example 2.11, the value of absence is changed from one to zero for the first record when the statement IF ABSENCES=1 THEN ABSENCES=0 is executed.❶

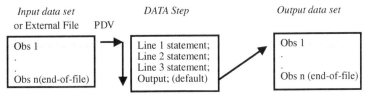

Input data set   *DATA Step*   *Output data set*
or External File   PDV

Obs 1 → Line 1 statement; → Obs 1
. Line 2 statement; .
. Line 3 statement; .
Obs n(end-of-file) Output; (default) Obs n (end-of-file)

tim lou 1 6.0 3 10/23/78   if absences=1 then absences=0;   tim lou 0 6.0 3 6870
/*output statement (default)*/

PROGRAM DATA VECTOR

NAME	NAME	ABSENCES	QUIZ	CLASS	DOB	_N_	_ERROR_
TYPE	$	N	N	N	N	N	N
LENGTH	8	8	8	8	8	8	8
VALUE	tim lou	0	6.0	3	6870	1	.

4. Write the values from the PDV to the output data set. After reaching the last DATA step statement, the values of each variable are written to the data set. The OUTPUT statement is the last implied statement in the DATA step. If the OUPUT statement is explicitly specified prior to the end of the DATA step, then the observation is written to the output data set at that time and not at the end of the DATA step.

DSNAME: First record

NAME	NAME	ABSENCES	QUIZ	CLASS	DOB
TYPE	$	N	N	N	N
LENGTH	8	8	8	8	8
VALUE	tim lou	0	6.0	3	6870

5. Loop back to the top of the input data set or external file.
6. Read in the next sequential observation from the input data set or the external file. **Note that SAS sees and processes only one observation at a time.**

DSNAME: Second record

NAME	NAME	ABSENCES	QUIZ	CLASS	DOB
TYPE	$	N	N	N	N
LENGTH	8	8	8	8	8
VALUE	sally	2	10.1	2	6576

7. Continue to execute steps 1–6 until the end-of-file marker is reached in the input data set or the external file.
8. Close the output data set. All observations have been processed and saved. The data set has now been created and is ready to be used by another DATA step or any SAS procedure.

Example 2.12 Create data set from reading data from another SAS data set.

```
data dsname;

  set class;

/* PDV - SAS guts */
if absences = 1 then absences = 0;
/* output statement (default) */
run;
```

There are two main differences when using the SET statement instead of the INPUT statement as far as the compile and execution process. In the compile phase steps 3 and 4, when using the SET statement, **SAS reads the descriptor portion of the input data set and places all variables directly in the PDV without using the input buffer.** In the execution phase step 2, when using the SET statement, SAS reads observations sequentially from the input data set instead of initializing all variables to missing before each execution of the DATA step. Thus, the data values from the input data set are read into the PDV. Any new variable in the new data set will be placed in the PDV and initialized to missing. Another variation of creating data sets from a data set specified with multiple SET statements is seen in Example 2.13.

**Example 2.13   Create data set from reading data from another
                SAS data set within a DO loop.**

```
data A;
  input patno source $ gender $;
  Cards;
1 A male
3 A male
;
run;


title 'DSNAME1: data set without an unconditional ELSE
clause';
data dsname1;
  do until (eof=1);   ❶
     set A end=eof;
     if patno <= 2 then group = 'Early';   ❷
     output;
   end;
run;


proc print data=dsname1;
  var patno group;
run;


title 'DSNAME2: data set with an unconditional ELSE
clause';
data dsname2;
  do until (eof=1);
     set A end=eof;
```

```
    if patno <= 2 then group = 'Early';
    else group = 'Late'; ❸
    output;
  end;
run;


proc print data=dsname2;
  var patno group;
run;
```

Output

```
DSNAME1: data set without an unconditional ELSE clause
```

Obs	patno	group
1	1	Early
2	3	Early ❹

```
DSNAME2: data set with an unconditional ELSE clause
```

Obs	patno	group
1	1	Early
2	3	Late ❺

In Example 2.13, there is a SET statement within a DO UNTIL loop. This code will read each record in data set A until the end-of-file marker is read ❶. When creating the first data set, dsname1, there is an IF–THEN statement without an unconditional ELSE clause ❷. When creating the second data set, dsname2, an unconditional ELSE clause is included in the code ❸. Because the SET statement executes multiple times within a DO loop, SAS does not initialize new variables to missing. The values of the new variable are retained unless changed. This can be seen as the difference between data sets dsname1 and dsname2. The value of group in the second record is retained from the first record in data set dsname1 because there was not an unconditional ELSE clause to change the group value ❹. It is better to specify an unconditional ELSE clause to assign the value of the group variable correctly for the second record in data set dsname2 ❺. See Example 1.17 for more information on using multiple SET statements.

See the following SAS papers for more information about the compile and execute steps: *"Data Step Internals: Compile and Execute," "The Secret Life of the DATA Step."*

Chapter 2. Creating Data Structures—Chapter Summary

Creating Temporary and Permanent SAS Data Sets

Example	Description
DATA DEMOG MYLIB.CENTER; SET DSNAME; RUN;	Creates temporary data set DEMOG and permanent data set CENTER from DSNAME data set.

Conditional Assignments and Conditional Subsets

```
If absences=2 then class = 4;    /* change value based on
                                    variables value */


select(test_score);
   when(10) grade = 'A';
   when(9) grade = 'B';
   otherwise grade = '?';
end;
set formatted (where=(name='sally')); /* subset data set
                                         based on condition */


if absences = 2;
```

Drop and Keep Statements/Options

```
data step statements - when writing to a SAS data set:
DROP quiz class;
KEEP name dob;


data set options - when reading from input writing to
output SAS data set:
(DROP = quiz class)
(KEEP = name dob)
```

Differences between WHERE and IF Conditions to Subset Data Sets

WHERE Condition	IF Condition
Process data in data sets.	Process raw data and data sets.
Requires SET, MERGE, or UPDATE statement if within DATA step because must process data sets.	Can use clause in any DATA step including, for example, INPUT statement.
Cannot use any automatic variables such as _N_ in condition.	Can use automatic variables such as _N_ in condition.
Cannot apply new variables in condition.	Can apply new variables in condition.
Is not compatible with data set options: OBS=, POINT=, FIRSTOBS=.	Is compatible with data set options: OBS=, POINT=, FIRSTOBS=.
Applies condition before merging data sets.	Applies condition after merging data sets.
Can use special operators in condition.	Cannot use special operators in condition.
Can apply directly to SAS procedures.	Cannot apply directly to SAS procedures.
More efficient method.	Less efficient method.
Uses index if available.	Does not use index.

Creating and Manipulating SAS Date Values

Example	Description
`newdt = 01jan2002d;`	`/* create date variable and assign date value */`
`If dob gt 02feb1978d;`	`/* condition on date variable */`
`Format dob newdt mmddyy8.;`	`/* format date variable */`

When you use an SAS informat to read a date, SAS software converts it to a numeric **date value.** An SAS date value is the number of days from January 1, 1960, to the given date.

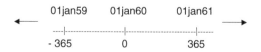

Using DATA Step Statements to Export (or Write) Data

```
filename outfile output_file.txt;

data _null_;
  file outfile;
  set formatted;
  put name = name $ 8-15 birth date = dob date9. +2
quiz 4.1 +2 class=;
run;
```

Key Compile and Execute Steps

	Description	SAS Statements
1. Compile Phase	Checks syntax; error message if it fails. Converts code to machine language. Establishes definition of input and output files.	Location relevant: ARRAY, ATTRIB, BY, FORMAT, INFORMAT, LENGTH, WHERE
	DATA step processing. Create the following: input buffer, program data vector (PDV), and descriptor information.	Location is irrelevant: DROP, KEEP, LABEL, RENAME, RETAIN, TITLE, OPTIONS
	Compile time SAS statements. Run in any order, define for all variables: type and length, cannot be used in conditional executions.	* For character variables, *first* LENGTH or ATTRIB is used * For numeric variables, *last* LENGTH or ATTRIB is used
	For each input record, PDV processes _N_ and _ERROR_ automatic variables.	
2. Execution Phase	DATA step processing of execution statements: The DATA statement, which names the data set, is the first statement to execute. Initializes each variable in the input buffer to missing. Variables from the SET statement data set bypass the input buffer and are placed directly in the PDV.	Almost all other SAS statements not listed as compile-time statements, including: ASSIGNMENTS, IF–THEN, SELECT–WHEN, FUNCTIONS, DO loops INFILE, FILE

(continued)

Key Compile and Execute Steps (*continued*)

	Description	SAS Statements
2. Execution Phase	For each input record, process all statements in order of sequence within the DATA step. All records are processed unless conditional execution is applied. Write input record to output data set. Loop to next input record and repeat iteration process to execute group of statements for each input record. Run-time errors result from programming errors, logic errors, or data issues. Automatic variables _N_ and _ERROR_ track records of data errors. Data errors result when input data does not confirm with what SAS expects; e.g., variable type conversion or missing values may be assigned.	* Executed in order of sequence. Statements correspond to specific actions that are part of the observational loop in a DATA step.

Chapter 2. Creating Data Structures—Chapter Questions

Question 1: How do you assign a date constant of January 1, 2002?

Question 2: What is the order of sequence if both DROP and KEEP statements are applied within the DATA step?

Question 3: During the execution phase, does the input buffer always store the values of the record?

Question 4: What are at least two methods to control the selection of observations within a data set?

Question 5: Is the INFORMAT statement a compile-time statement?

Question 6: Can formats be used in PUT statements within a DATA step to write to an external file?

Question 7: What is one method for applying a subset condition and restricting the number of observations displayed when using the PRINT procedure? For example, how can you list only the male patients within the first 20 records?

Question 8: When creating SAS data sets from external files, how does SAS determine the order of the variables created?

Question 9: In which of the two phases, compile or execute, are data values written to the output data set?

Question 10: What is the advantage of creating a permanent data set instead of accessing a temporary data set and how is it referenced?

Question 11: What is the SAS data value for the date January 1, 1960, and what type of variable stores date values?

Question 12: What is the sequence of statements in terms of compile-time and execution-time statements as they are executed within a DATA step?

Question 13: When using the IF statement to change a variable's value conditionally, does the change take place in the input buffer or the PDV?

Question 14: What two statements are required to write to an external file from a DATA step?

Question 15: Are the DROP and KEEP statements used to control selection of variables or observations within a data set?

Question 16: What is the following variable list shortcut equivalent to? VAR ROOT1–ROOT3;

Question 17: When creating data sets, if the OUTPUT statement is not specified, is there an implied OUTPUT statement? If so, where is the implied OUTPUT statement located in the DATA step?

Question 18: Which WHERE condition would be executed if you specified a WHERE data set option along with a WHERE statement in the same DATA step?

Question 19: In which of the two phases, compile or execute, is the descriptor portion of the output data set created?

Question 20: Is it possible to execute the following code without getting an error message?

```
data test;
  if sex = 'Male' then drop age;
run;
```

Question 21: If the compilation phase results in an error, is the data set created?

Question 22: What condition can be applied to process the first record in the data set?

Question 23: Is there a difference between specifying the DROP= data set option before the OBS= data set option on the same data set?

Question 24: Can a WHERE statement include new variables or temporary variables such as _N_?

Question 25: Can a WHERE statement be applied to DATA steps with an INPUT statement?

Question 26: If the YEARCUTOFF option is set to 1920, what is the 100-year span window SAS uses to read dates with only two-digit years?

Chapter 3

Managing Data

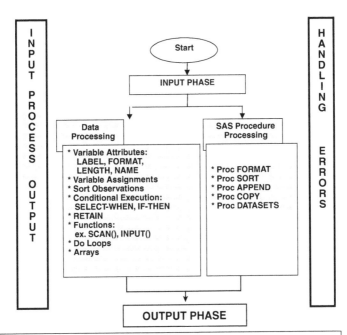

| Start |
| INPUT PHASE |

Data Processing
* Variable Attributes:
  LABEL, FORMAT,
  LENGTH, NAME
* Variable Assignments
* Sort Observations
* Conditional Execution:
  SELECT-WHEN, IF-THEN
* RETAIN
* Functions:
  ex. SCAN(), INPUT()
* Do Loops
* Arrays

SAS Procedure Processing
* Proc FORMAT
* Proc SORT
* Proc APPEND
* Proc COPY
* Proc DATASETS

| OUTPUT PHASE |

I N P U T   P R O C E S S   O U T P U T

H A N D L I N G   E R R O R S

PROCESS FLOW DIAGRAM

PHASE	CHAPTER. DESCRIPTION
INPUT:	1. Accessing Data
	2. Creating Data Structures
PROCESS:	3. Managing and Summarizing Data
OUTPUT:	4. Generating Reports
HANDLING ERROR:	5. Diagnosing and Correcting Errors
V8.2/9.1:	6. Integrity Constraints, Generation DS, Audit Trials

3.1 Introduction

This chapter focuses on how to manage your data once your data set is created. You will discover how to modify the attributes of your variables and the various ways to assign values. You will also be introduced to code-saving methods of programming such as DO loops, SAS arrays, and SAS functions. The last section of this chapter discusses several utility SAS procedures to investigate SAS data libraries. It is not within the scope of this book to show all of the options available in these SAS procedures.

3.2 Modifying Variable Attributes in the Data Set

3.2.1 General Rules

Variable attributes, such as label, format, and length, for both character and numeric variables can be assigned once the variable has been created. Modifications to variable attributes can be made using the same SAS statements. If the attributes are modified in a DATA step, they become **permanent.** If attributes are modified in a SAS procedure, such as with a LABEL or FORMAT statement, then the change is only temporary and applies only for the duration of the SAS procedure.

3.2.2 Applying Labels to Variables in an SAS Data Set

Syntax:
LABEL <variable> = 'label';
　(OR)
ATTRIBUTE <variable>　LABEL='label';

The LABEL statement applies a descriptive label to a variable. The descriptive label not only serves to document the variable, but can also be used as a column header in a report. **The maximum length a label can be is 256 characters.** In a LABEL statement, the variable name is listed followed by the equals sign and the descriptive label in quotes. Multiple variable names and their labels can be listed in the LABEL statement, separated by spaces. In addition to the LABEL statement, the ATTRIBUTE statement can also be used to apply a label to a variable. The ATTRIBUTE or ATTRIB statement is similar in syntax to the LABEL statement. The keyword ATTRIB is followed by the variable name. The label is then specified in quotes after the LABEL= statement. Later in this chapter, you will see that the ATTRIB statement can also be used to assign variable formats and lengths.

Example 3.1 Applying labels to variables in a SAS data set.

```
data patno;
 label patno='Patient Number';  ❶
 attrib gender label='Gender (male, female)';

 input patno source $ gender $;
 cards;
1 C male
2 C female
3 C male
5 C male
;
run;

proc contents data=patno;❷
run;

proc print data=patno label;❸
run;
```

Output

```
                    The CONTENTS Procedure


-----Alphabetic List of Variables and Attributes-----


#     Variable   Type   Len   Pos   Label


2      gender    Char    8     8    Gender(male, female)❷
1      patno     Num     8     0    Patient Number
3      source    Char    8    16
                               ❸
```

	❸	Gender (male, female)	source
Obs	Patient Number		
1	1	male	C
2	2	female	C
3	3	male	C
4	5	male	C

In Example 3.1, the *patno* variable has the label applied using the LABEL statement ❶ while the *gender* variable has the label applied using the ATTRIBUTE or ATTRIB statement. The *gender* label helps to document the valid values of male and female for that variable. ❷ In a DATA step, multiple LABEL statements or ATTRIB statements can be specified or all of the variable labels can be applied in one SAS statement. The advantage of the ATTRIB statement over the LABEL statement is that with the ATTRIB statement, multiple attributes can be assigned. As can be seen, the PRINT procedure with the LABEL option uses the label as column headers for the *gender* and *patno* variables.❸ This option will be discussed in Chapter 4.

3.2.3 Applying Formats to Variables in an SAS Data Set

Syntax:
FORMAT <variable> format;
  (OR)
ATTRIBUTE <variable> FORMAT=format;

Formats were first introduced in the FORMATTED INPUT statement section of Chapter 1. **Formats used in the INPUT statement (informats) are specified to read the data correctly, while format statements (formats) are specified to display data correctly.** In general, formats and informats are similar in syntax and must contain a period.

A format can be applied to a variable in order to specify the way that the variable is displayed without changing the values that are actually stored in the data set. SAS provides various formats for numeric data, character data, and dates and times. When using a FORMAT statement, the variable name is listed, followed by the format. The format specified must match the variable type (numeric or character). If it does not, SAS will issue an ERROR message. Like the LABEL statement, multiple variables can have their formats assigned in the same SAS statement. If multiple variables need to be assigned the same format, all of the variables can be listed before the format name. In addition to the SAS formats, a variable can be assigned a user-defined format. Later in this chapter, you will see how to create user-defined formats. The ATTRIB statement can also be used to assign a format to a variable using the same type of syntax that is used to assign a label. The difference is that following the variable name is FORMAT= followed by the format name. If the ATTRIB statement was already used to define another attribute, the format can be assigned by adding FORMAT= followed by the format name. Note that specifying a FORMAT statement in SAS procedures temporarily overrides the variable's permanent format when the data values are displayed.

Example 3.2   Applying formats to variables in an SAS data set.

```
data dob;
  attrib dob label='Date of Birth'; ❷
  format dob date9.; ❶

input patno source $ gender $ dob mmddyy10.;
  cards;
1 C male 01/12/2002
2 C female 11/14/1998
3 C male 07/22/1996
5 C male 05/23/1995
;
run;

proc print data=dob label; ❸
run;

proc contents data=dob;
run;
```

Output

```
        Date of
Obs     Birth     patno   source   gender
 1     12JAN2002    1       C       male
 2     14NOV1998    2       C       female
 3     22JUL1996    3       C       male
 4     23MAY1995    5       C       male


            The CONTENTS Procedure
-----Alphabetic List of Variables and Attributes-----

 #     Variable     Type    Len Pos Format    Label

 1      dob         Num      8   0  DATE9.    Date of Birth
 4      gender      Char     8  24
 2      patno       Num      8   8
 3      source      Char     8  16
```

In Example 3.2, the DATE9. format is applied to the *dob* variable. ❶ Without the format, the *dob* variable would be displayed as the number of days from January 1, 1960 and would be difficult to understand. Since the

ATTRIB statement was used, both the format and the label are being applied in the same statement.❷ In order to use the label as a column header in the output of the PRINT procedure, the LABEL option was used.❸ (This option will be discussed in Chapter 4.) In the output for the CONTENTS procedure, the format and the label for the *dob* variable are displayed.

3.2.4 Explicitly Defining the Lengths of Variables

Syntax:
LENGTH <variable> length;
  (OR)
ATTRIBUTE <variable> LENGTH=length;

The LENGTH statement is used to assign the space allowed to store data values and to define whether a variable is numeric or character. In Version 8.2, character variables can have a length of 1 to 32,767 and numeric variables can have a length of 3 to 8 or 2 to 8, depending on the host system. When assigning a length to a variable, it is important to make sure that the length is long enough or you may accidentally truncate the data. For character variables, ensure that the length is larger than the maximum number of raw columns used. For numeric variables, the number of raw columns does not correspond to the variable's length.

 Since the variable length is defined in the compilation phase, it is important to specify the LENGTH statement early in the DATA step in order to make sure that the correct length is assigned. In general, when creating new variables, the length of the variable is determined by the first reference within the program. But if the LENGTH or ATTRIB statement is processed first, then SAS does not use the variable's first value in the data file or data set to determine the variable's length automatically.

Example 3.3 Explicitly defining the lengths
              of character variables.

```
data dob;
 attrib source length=$1
        gender length=$5;  ❶

 input patno source $ gender $ dob mmddyy10.;
 cards;
1 C male    01/12/2002
2 C female 11/14/1998
3 C male    07/22/1996
5 C male    05/23/1995
;
run;
```

```
proc print data=dob;
run;

proc contents data=dob;
run;
```

Output

Obs	source	gender	patno	dob
1	C	male	1	15352
2	C	femal ❷	2	14197
3	C	male	3	13352
4	C	male	5	12926

                The CONTENTS Procedure

-----Alphabetic List of Variables and Attributes-----

#	Variable	Type	Len		Pos
4	dob	Num	8		8
2	gender	Char	5		17
3	patno	Num	8		0
1	source	Char	1	❸	16

In Example 3.3, the ATTRIB statement was used to specify the length of both the *gender* and *source* variables. ❶ If the ATTRIB statement was not specified, then the default length of each character variable would be 8 because of the LIST INPUT statement. For new variables not specified in the ATTRIB, LENGTH or INPUT statement, SAS uses the variable's first value in the program to determine the length. See Example 5.17 on how to avoid character field truncation. For the *gender* variable, the length was specified as 5. Looking at the output from the PRINT procedure, you can see that the value for female was truncated since the value for the LENGTH statement was too short. ❷ When specifying the LENGTH statement, you need to make sure that it is long enough to capture all of the data. This is easy to see with character variables, but with numeric variables the precision of the variable is closely tied to the length of the variable. Extreme caution should be used when specifying the length of numeric variables that may contain fractions, as it will affect the precision of the variable. The length was also specified for the *source* variable as 1. ❸ If no length is specified, then by default, SAS assigns a length of 8. The *source* variable shows an advantage of using the LENGTH statement in that by specifying a shorter length you are able to save disk space. The valid values that can be used in the LENGTH statement for numeric variables are system dependent, so the SAS documentation for your operating system should be consulted.

3.2.5 Modifying Variable Names

Syntax:
RENAME <old name> = <new name>;

The purpose of the RENAME statement is to change a variable's name; all other attributes of the variable remain the same. In the RENAME statement, the new variable name is specified after the equal sign, which comes after the original variable name. The RENAME statement can be used either as a data set option or as a DATA step statement. **The difference between the RENAME statement and the RENAME option is that the RENAME statement will apply the change to all output data sets created.** If you want to name variables differently in different data sets, then you must use the RENAME data set option. The other difference is that the RENAME statement cannot be used in SAS procedures. Note that caution must be used when applying the RENAME statement or option since SAS will no longer save the original variable name in the new data set. **In addition, because SAS does not rename the variable until it writes to the output data set, the original variable name can still be referenced within the DATA step.** The new variable name only exists in the new data set. Note that in Version 8.2, it is possible to take advantage of making the variable names case-sensitive. See Chapter 6, Section 6.2 for more information.

Example 3.4 Modifying variable names.

```
data dob ;
  input patno source $ gender $ dob mmddyy10.;
  rename gender = sex; ❶
  cards;
1 C male 01/12/2002
2 C female 11/14/1998
3 C male 07/22/1996
5 C male 05/23/1995
;
run;

proc print data=dob;
run;
```

Output

Obs	patno	source	sex	dob
1	1	C	male	15352
2	2	C	female	14197
3	3	C	male	13352
4	5	C	male	12926

Example 3.4 is simple: you can see that with the RENAME statement, the variable *gender* is renamed *sex*. ❶ Remember that the new variable will still have all of the original variable's attributes. In this example, since there is only one output data set, the RENAME data set option would produce the same result. See Table 2.1 for more information on data set options.

3.2.6 Creating and Using User-Defined Formats

Syntax:
LIBRARY <library reference> '<physical file location>';
PROC FORMAT <LIBRARY = library reference> FMTLIB;
   VALUE <format name> <value1>='format label1'
                         <valuen>='format labeln';

RUN;

FORMAT <variable> <$> <format name>.;

FORMAT Catalog

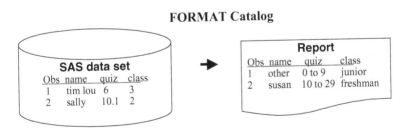

The FORMAT procedure is used to create user-defined formats. User-defined formats are formats in which the user creates the values that will be displayed when the format is applied. The main reason to create a format is to display the data in a specific way. Formats can be created for both numeric and character variables. There are a few rules that must be followed when creating formats using the VALUE statement. **The name of the format must be less than or equal to 8 characters for numeric format names. Character format names must be less than or equal to 7 characters, since SAS requires that a "$" be used in the first position to identify that the format is a character format.** When creating the format value that will be associated with one or more variables, enclose character format values, but not numeric format values, in quotes. The format label must be enclosed in quotes and must be less than 200 characters. When the FORMAT procedure is used, the formats created can be either permanent or temporary. In order to make the formats permanent, the LIBRARY= option must be used in the PROC FORMAT statement. Permanent formats can be accessed by using the LIBRARY libref without needing to recreate the format catalog. A useful option to specify is FMTLIB, which displays the formats saved in the format library.

After a format has been created with the FORMAT procedure, you are ready to use it. In order to do this, the format must be applied to a variable. This is done using the FORMAT or ATTRIB statement in the DATA step or in an SAS procedure. If the format is assigned in a DATA step, the format is permanently assigned to the variable. If the format is assigned in an SAS procedure, it will be assigned only for the duration of that procedure. See the SAS paper *Using and Understanding SAS Formats* for more information on formats.

Example 3.5   Creating and using user-defined formats.

```
data class3;
   input name $8. @10 absences 2. @15 quiz 4.1 @21 class
   1. @23 dob mmddyy8.;
cards;
tim lou   1       6.0    3 10/23/78
sally     2      10.1    2 01/02/78
john      2       8.0    3 06/22/78
;
run;
```

❺

```
proc format fmtlib;
  /* character format name  - $namefmt.*/
   value $namefmt   'sally' = 'susan'   /* value - sally,
                                           label - susan */
             ❷           other = 'other';
```

```
/* numeric format name  - CLASSFMT */
   value classfmt   2 = 'FRESHMAN'       /* value - 2,label-
                                              FRESHMAN */
                    3 = 'JUNIOR'         /* value - 3, label
                                            - JUNIOR */
                 other = 'OTHER';
```

```
/* numeric format using range of values */
```

```
value qsgrp      0 - 9 = '0 TO 9'
        ❶       10 - 29 = '10 TO 29';

/* numeric format using list of values */
value regfmt 1, 2, 3, 4 = 'NW'
                  5, 6 = 'NE';
quit;

data class_fmt;
  set class3;
                  /* date9. - SAS-supplied format */
  format name $namefmt. class classfmt. quiz qsgrp. dob
  date9.;  ❸
run;

proc print data=class_fmt;
run;
```

Output

❻

FORMAT NAM E: CLASSFMT LENGTH: 8 NUMBER OF VALUES: 3		
MIN LENGTH: 1 MAX LENGTH: 40 DEFAULT LENGTH 8 FUZZ: STD		
START	END	LABEL (VER.8.2 15AUG2004:08:00:05)
2	2	FRESHMAN
3	3	JUNIOR
OTHER	**OTHER**	OTHER

FORMAT NAME: QSGRP LENGTH: 8 NUMBER OF VALUES: 2		
MIN LENGTH: 1 MAX LENGTH: 40 DEFAULT LENGTH 8 FUZZ: STD		
START	END	LABEL (VER.8.2 15AUG2004:08:00:05)
0	9	0 TO 9
10	29	10 TO 29

FORMAT NAME: REGFMT LENGTH: 2 NUMBER OF VALUES: 6		
MIN LENGTH: 1 MAX LENGTH: 40 DEFAULT LENGTH 2 FUZZ: STD		
START	END	LABEL (VER.8.2 15AUG2004:08:00:05)
1	1	NW
2	2	NW
3	3	NW
4	4	NW
5	5	NE
6	6	NE

```
FORMAT NAME: $NAMEFMT LENGTH: 5 NUMBER OF VALUES: 2
MIN LENGTH: 1 MAX LENGTH: 40 DEFAULT LENGTH 5 FUZZ: 0
```

START	END	LABEL (VER.8.2 15AUG2004:08:00:05)
sally	sally	susan
OTHER	**OTHER**	other

	Obs	name	absences	quiz			class	dob
❹	1	other	1	0	TO	9	JUNIOR	23OCT1978
	2	susan	2	10	TO	29	FRESHMAN	02JAN1978
	3	other	2	0	TO	9	JUNIOR	22JUN1978

In Example 3.5, several types of formats have been created. Note that formats such as GSGRP and REGFMT use a range of values or a list of values to assign a single label ❶. The $ NAMEFMT is a character format that reassigns a name ❷. Notice that the format name is only 7 characters and that there is a "$" before the format name. Also, the keyword OTHER is used in the format, which will assign any other name that has not been specified as a value. Additional keywords available with the FORMAT procedure include LOW and HIGH. Note that the LOW keyword does not include missing values. The CLASSFMT is a numeric format, so it can have a format name of 8 characters, since "$" is not used for numeric variables. The OTHER keyword is also used as a value for CLASFMT. In both cases, the keyword should not be enclosed in quotes. The QSGRP format is an example of how to assign a format where a range of values is used. In the format, if the value of the variable is within the range specified, then that is the format value that will be applied. This is similar to the REGFMT format where a list of values is used. Similarly, if the value of the variable is in the list, the label will be assigned as the format. Note that user-defined formats can be treated just like SAS formats in the FORMAT statement ❸. In the output, you can see how the values are displayed when the formats are applied ❹. Note that the data values did not change. With the FMTLIB option ❺, all formats created are displayed ❻.

3.3 Assignment Statements and Conditionally Executed SAS Statements

3.3.1 General Rules

Syntax:

Assignment Statements:
Variable1= expression;
Variable2='expression';

If (expression1) then do;
end;

Mutually Exclusive Conditional Statements;
If (expression1) then (expression2);
Else if (expression1) then (expression2);
Else (expression2);

Select (variable);
 When (expression1) expression2;
 When (expression1) expression2;
 Otherwise expression2;
End;

Assignment statements are used to assign values to new variables and to modify values of existing variables. Assignment statements can be specified with or without a condition. In addition, multiple statements can be executed sequentially when a condition is met. As seen in Section 2.3, SAS enables you to assign variables conditionally, based on whether a certain condition is met. Both numeric and character variables can be created or modified.

The components of a conditionally executed statement can be broken into two parts: condition and execution. Since the execution statement will only process if the condition is met, it is recommended that you test the condition to ensure that the right data is captured. Once the condition component has been confirmed, then the execution component should be tested. This can be done without the condition to ensure correct results. After you test each component, the combined parts should work correctly. It is recommended that you include the ELSE and OTHERWISE statements in the IF–THEN and SELECT–WHEN statements respectively, to ensure mutually exclusive conditions. In this way a value will always be assigned to a new variable and the "Uninitialized Variable" message in the SAS log can be avoided. Table 3.1 shows the testing of each component of the conditional execution statement.

Conditional statements are used to execute selected statements conditionally. The condition must match the type of variable that is being compared. If your conditioning or assignment variable is numeric, a numeric value must appear to the right of the equal sign. Likewise, if a character variable is being compared or assigned, the value to the right of the equals sign must be a character value that appears in quotes. Text that appears in conditions on character variables is case sensitive; e.g, there is a difference between "Male" and "male." Make sure to have balanced quotation marks to prevent an error message. When making a comparison, any of the following operators are valid: EQ (equal), NE (not equal), GT (greater than), LT (less than), GE (greater than or equal), and LE (less than or equal).

Table 3.1 Testing of Conditional Execution Statements

Test Condition Component	Test Execution Component
Do you have the condition, variable, and value required to catch your data?	Do you have the correct assignment statement, variable, value, or expression?
Only when condition is true will the assignment statement execute.	The execution statement is generally an executable statement.

Example 3.6 Assignment statements and conditionally excuted statements.

```
data class3;
  input name $8. @10 absences 2. @15 quiz 4.1 @21 class
  1. @23 dob mmddyy8.;
cards;
tim lou   1      6.0    3 10/23/78
sally     2     10.1    2 01/02/78
john      2      8.0    3 06/22/78
;
run;


data condition;
 set class3;

 attrib name label='Student Name' format=$8.;
 format dob newdt date9.;

 if absences = 2 then do; ❶
  class = 4; /*one or more statements */
  newdt = '01jan2002'd;
end;
select(name);            ❷
 when('tim lou') name = 'Tim Lou';
 when('sally') name = 'Sally';
 otherwise name = 'Unknown';
end;

run;


proc print data=condition;
run;
```

Output

Obs	name	absences	quiz	class	dob	❹	newdt
1	Tim Lou ❸	1	6.0	3	23OCT1978		.
2	Sally	2	10.1	4	02JAN1978		01JAN2002
3	Unknown	2	8.0	4	22JUN1978		01JAN2002

The first conditional statement in Example 3.6 is used to assign values to a variable if the condition is met ❶. Notice that since the expression must match the variable that it is being set to, the *newdt* expression must be in the form of 'DDMMMYYYY'd or 'DDMMMYY'd. The second conditional statement is in the form of a SELECT–WHEN statement ❷. After the SELECT statement, WHEN statements are used to set the value of the *name* variable based on the conditional criteria. The OTHERWISE statement is used to set all values of the variable that have not matched the other WHEN statements. Results of the conditional executed statements on the *name* variable can be seen from the PRINT procedure ❸. Also, class = 4 and newdt = 01JAN2002 when absences = 2. ❹

3.4 Using the Retain Statement to Accumulate Variable Values Across Executions of the DATA Step

3.4.1 General Rules

With each iteration of the DATA step, by default, all variables are initialized to missing. If the variable is not assigned an initial value, the output data set will have the variable set to missing. You can use the RETAIN statement to initialize a numeric or character variable to a specific value. Also, with the RETAIN statement, a variable that is assigned a value with an INPUT or assignment statement will retain its value from the current iteration of the DATA step to the next record.

In general, the RETAIN statement is used on new variables in the DATA step. **The RETAIN statement preserves the variable's value with each iteration of the DATA step.** By default, without a RETAIN statement, the variable is automatically set to missing with each DATA step iteration. SAS initializes the retained variables to missing before the first execution of the DATA step if you do not supply an initial value. The RETAIN statement is not an executable statement; it is a compilation-time-only statement that creates variables if they do not already exist. See Example 2.11 and Example 2.12 in Chapter 2 for more information on the compilation and execution phases.

3.4.2 Using the Retain Statement to Accumulate Variable Values Across Executions of the DATA Step

Syntax:

RETAIN <variable> <initial value>;

The RETAIN statement is usually used for counting values or summing values across iterations of the DATA step. The RETAIN <variable name> 0; statement is required for accumulating totals. If you are counting values across iterations of the DATA step, an assignment statement such as <variable name> = <variable name> +1; is needed to increment the count variable. If you are summing variables across iterations of the DATA step, an assignment statement such as <variable name = <variable 1> + < variable 2> or the SUM( ) function is needed. The SUM( ) function is a better choice, since it will sum all of the nonmissing values and not cause a message to be written to the SAS log about missing values.

Example 3.7   Using the RETAIN statement to accumulate totals.

```
data tests;
   input name $ test_score;
   cards;
   Tim  9
   Tim  7
   Tim  7
   Sally 10
   Sally 7
   John  8
   John  7
   John  6
;
run;


proc sort data=tests;
   by name; ❸
run;

data grades;
   set tests;
   by name;
```

```
    retain total count city 'Greenville';  ❶
    format average 4.1;
if first.name then do;
      total=0;  ❷
      count=0;
    end;
    count=count + 1;
    total=sum(total,test_score);  ❹

    if last.name then do;
      average=total/count;  ❺
      output;
    end;

run;

proc print data=grades;
run;
```

Output

Obs	name	test_score	total	count	❻ city	average
1	John	6	21	3	Greenville	7.0
2	**Sally**	7	17	2	**Greenville**	8.5
3	Tim	7	23	3	Greenville	7.7

In Example 3.7, you want the output data set to be one record per subject where the average test score is calculated. In order to do this, you use the RETAIN statement to create the *total* and *count* variables. ❶ Note that character variables can also be returned, e.g., city will keep the value Greenville until changed. The *count* will be the number of tests and the *total* will be the total of the test scores. The RETAIN statement preserves the values of the variables over iterations of the DATA step. The values of the variables are initialized in the FIRST.name statement instead of in the RETAIN statement, since we want to start each variable at zero for each name. ❷ In order to use the FIRST.name and LAST.name statements we have to sort the input data set by the BY variable name. ❸ The *count* and *total* variables are then increased with each iteration of the DATA step until the LAST.name record is reached. ❹ When the LAST.name record is reached, the average is calculated for each *name* and that record is written to the data set.❺

Note that an implied RETAIN statement is used when specifying a SAS statement in the form <variable> + <number> ; or <variable 1> + <variable 2>;.

The first SAS statement, <variable> + <number> ;, is equivalent to the following two SAS statements: RETAIN <variable> 0; <variable> = SUM(<variable> + <number>);.

The second SAS statement, <variable 1> + <variable 2>;, is equivalent to the following two SAS statements: RETAIN <variable 1> 0; <variable 1> = SUM(<variable 1> + <variable 2>);.

3.5 Using SAS Functions to Manipulate Character Data, Numeric Data, and SAS Date Values

3.5.1 General Rules

Over 300 different SAS functions exist for manipulating character data, numeric data, or date values. The *SAS Language Reference* manual describes the use of all the different functions. **SAS functions are calculations or transformations of values that are preprogrammed by the SAS Institute.** For most SAS functions, you will need to supply arguments, but a few will obtain their arguments directly from the operating system. The syntax of a SAS function is the function name with the arguments enclosed in parentheses after the function name. Parentheses must always be supplied even if the function does not require an argument. Note that functions can process multiple variables across one observation as compared to SAS procedures, which process only one variable per observation. SAS enables the use of multiple nested functions to be specified for combined effects. Always work from the inside expression to the outer expression when resolving nested functions. Note that in Version 9.1, there are many new functions that permit easier manipulation of data. In some cases, new Version 9.1 functions replace the expression of two older functions. See Chapter 6, Section 6.3 for more information. See also the following SAS papers for information on SAS functions: *Discovering the FUN in SAS Functions, Programming with SAS Functions, with a Special Emphasis on SAS Dates and Times*.

3.5.2 Using SAS Functions to Manipulate Character Data, Numeric Data, and SAS Date Values

In general, character functions return a string value. There are several functions, such as the INDEX() and LENGTH() functions, however, that return a numeric value. There are a few things to keep in mind with functions that manipulate character data. Each character in a character string will take up one byte of memory. The maximum number of bytes that a character variable, character constant, or character expression can have is 32,767. So one concern when working with characters is that you do not go over the 32,767 byte limit. **A character variable always has a fixed length that does not change during a DATA step or by the length specified in a LENGTH statement.** The length of the variable is determined by the context of the first appearance of the variable in the DATA step. If you want to concatenate character values together, the || operator can be used. Again, be careful not to go over the 32,767 byte limit. **When you use most character functions, such as SUBSTR(), LEFT(), or TRIM(), the function will return the same length that is specified in their arguments. Other functions such as the SCAN() and REPEAT() functions will return a length of 200.**

SAS provides a function called the LENGTH() function to determine the length of strings stored in the variable. Note that this is not necessarily the length of the character variable, since the character data may contain blanks. **Thus, the string length may vary across records and will be less than or equal to the variable's length. The LENGTH() function will return the visual length of the string and not the length of the variable (memory length). The LENGTH() function will not include trailing blanks. If a null expression is used as the string (character string of length zero) the LENGTH() function will return a 1, not 0 as you might expect.**

Numeric functions return a numeric value and can take advantage of the OF keyword to include multiple variables without explicitly specifying the list of variables. For example, N(OF x1 – x3); is equivalent to N(x1, x2, x3);. The default length of numeric variables created from numeric functions is 8. **In general, it is recommended to include the number 0 in MEAN() and SUM() functions to assure the result is 0 if all argument variables have missing values.** Table 3.2 shows selected character, numeric, and date functions. The argument can be a variable, expression, or literal value.

Note that the SUBSTR() function is the only function that can be used at the beginning of an assignment statement with a value assigned to it. This is called a pseudo-variable. In the steps below the letter "D" is replaced by the letter "R" in the name variable.

- Original data: NAME = 'Don'
- DATA step statement: SUBSTR(NAME, 1, 1) = 'R'
- Updated data: NAME = 'Ron'

The TODAY() function is an example of a function that has no arguments, but the parentheses are still required. If a variable is created using a date function, the value that is returned may not look like a date unless it has been assigned a format.

Table 3.2 Selected Character, Numeric, and Date Functions

Function Call	Value Returned	Description
	Character Functions, e.g., (two spaces)**I**(one space)**live**(one space)**here**(three spaces)	
COMPRESS('I live here')	'I live here'	Removes all blanks.
INDEX(' I live here ', 'live') ---+---+---+-	5	Locates string 'live' at fifth position. Returns numeric value 5. Searches a character value; if found, it equals the number of first matched text, else 0.
INDEX(' I live here ', 'r ') ---+---+--+-	0	Does not locate string 'r ' because the letter 'e' exists after the letter 'r'. Returns numeric value 0.
INDEXC(' I live here ', 'rz') ----+----+--+-	12	Locates character string 'r' at the twelfth position. Returns numeric value 12.
LEFT(' I live here ') ----+----+----+-	'I live here' ---+---+---+-	Shifts string to left side. Moves two left blanks to right side.
LENGTH(' I live here ') --—+---+--—+-	13	Length of string is returned as numeric value. Trailing blanks are not counted.
RIGHT(' I live here ') ----+----+----+-	' I live here' ----+----+----+-	Shifts string to right side. Moves three right blanks to the left side.
SCAN(' I live here ', 1)	'I'	Returns first token string 'I'. Scan for words based on word position number, default delimiter = blank . < (+ & ! $ *) ; ^ - / , % |; Delimiters define the word separators in the string.
SCAN(' I live here ', -1)	'here'	Returns first token string counting backwards 'here'.
SUBSTR('New York, NY',5,4)	'York'	Extracts part of a value based on starting position 5 and length 4.
TRIM(' I live here ')	' I live here'	Removes right blanks.
UPCASE(' I live here ')	' I LIVE HERE '	Converts string to upper case.
	Numeric Functions	
N(1,2,.,.,3)	3	Returns the number of nonmissing values.
MIN(2,.,-3)	-3	Returns the smallest value of the nonmissing values.
MEAN(.,0,2,4)	2	Returns the mean of the arguments; missing values are excluded.

(continued)

SUM(.,0,2,4)	6	Returns the sum of the arguments; missing values are excluded.

Date Functions		
TODAY()	—	Returns the current date as a SAS date value. e.g., 16473
MONTH(startdt)	1–12	Returns the month from an SAS date value.
INTCK('WEEK',startdt, stopdt)	—	Returns number of weeks between startdt and stopdt. e.g., 3

Example 3.8   Using functions to manipulate data.

```
data class3;
  input name $8. @10 absences 2. @15 quiz 4.1 @21 class
1. @23 dob mmddyy8.;
cards;
tim lou    1   6.0  3 10/23/78
sally      2  10.1  2 01/02/78
john       2   8.0  3 06/22/78
;
run;

Data class_funct;
  Set class3;

  fname=upcase(scan(name,1));  ❶
  month=month(dob);
  format dob mmddyy8.;
run;

proc print data=class_funct;
run;
```

Output

```
                                              ❷
Obs    name     absences   quiz   class    dob      fname   month

 1    tim lou      1        6.0      3   10/23/78   TIM      10
 2    sally        2       10.1      2   01/02/78   SALLY     1
 3    john         2        8.0      3   06/22/78   JOHN      6
```

Example 3.8 shows the use of several functions. The *fname* variable is created by using the SCAN() and UPCASE() functions. ❶ The SCAN() function returns the first word in the *name* variable and then the UPCASE() function converts it to upper case. ❷ This is an example of specifying nested functions for a combined effect. The *month* variable is created by using the MONTH() function, which returns the month part of the *dob* date variable.

3.6 Use SAS Functions to Convert Character Data to Numeric Data and Vice Versa

3.6.1 General Rules

Because it is possible for numeric data to be stored in character variables, you should be careful when using character variables in mathematical operations or when using numeric variables in character string operations. If you try to do this conversion without the use of the PUT() or INPUT() functions, you will get a message in your SAS log that a numeric variable has been converted to a character variable or vice versa. **The PUT() and INPUT() functions are explicitly used to prevent messages in the SAS log when converting between variable types.**

3.6.2 PUT() and INPUT() Functions

Syntax:
<variable> = PUT(source, format);   /* Numeric to character */

<variable>= INPUT(source, informat);   /* Character to numeric */

The PUT() function can be used to convert numeric variables to character variables, while the INPUT() function can be used to convert character variables to numeric variables. The syntax between the two functions is similar; the function name is followed by the source, which can be an SAS variable or a constant whose value you wish to convert. **The PUT() function then has the format which you want applied to the variable. The format that you use must be of the same type as the source variable type.** Thus numeric variables require numeric formats and character variables require character formats. **The INPUT() function has the informat after the source variable. The informat is usually numeric and specifies the informat that is required in the new numeric variable.** In both the PUT() and INPUT() functions, informats and formats are required. **When using the PUT() function, the result is always a character string.** When using the INPUT() function, however, the result depends on the informat being numeric or character. Note that if a numeric variable is created, then the default length of the new variable is 8. Table 3.3 shows examples of the INPUT() and PUT() functions.

Without using the PUT() and INPUT() functions to control the conversion of variable types, SAS automatically converts the new variable as seen in Table 3.4. In addition, conversion notes will be placed in the SAS log to inform you of the automatic conversion. It is recommended to control this conversion to prevent this note.

Table 3.3 INPUT() and PUT() Functions

Function Call	Raw Value	Value Returned	Description
INPUT(agechar, 4.);	'30'	30	Converts character variable to numeric variable using numeric format.
INPUT(agechar, $4.);	'30'	' 30'	Formats character variable using character format.
PUT(age, 4.);	30	' 30'	Converts numeric variable to character variable using numeric format.
PUT(name, $7.);	'Smith'	'Smith'	Formats character variable using character format.

Table 3.4 SAS Automatic Conversion of Variable Types

Direction	Examples	Resolution
Numeric to Character	charvar = numeric value; charvar = charvar1 ‖ numeric value;	SAS will not automatically pad the converted numeric data value with zeros. Need to check logic.
Character to Numeric	numvar = numeric value in charvar; numvar = numvar1 + charvar;	SAS can truncate leading zeros from the character data values. May result in error if non-numeric values are in charvar so as to cause numvar to be set to missing

Example 3.9 Using the PUT() and INPUT() functions to convert data.

```
data class3;
  input name $8. @10 absences 2. @15 quiz 4.1 @21 chrclass
$1. @23 dob mmddyy8.;
  cards;
```

```
tim lou    1      6.0    3 10/23/78
sally      2     10.1    2 01/02/78
john       2      8.0    3 06/22/78
;
run;

data class_put;
  set class3;

  /* numeric dob saved as mmddyy format in character
variable*/
  chrdob = put(dob, mmddyy8.);  ❶
  /* character chrclass saved as 4. format in numeric
variable */
  numclass = input(chrclass, 4.);  ❷
  format dob date9.;
  Run;

  proc contents data=class_put;
  run;

  proc print data=class_put;
  run;
```

Output

The CONTENTS Procedure

-----Alphabetic List of Variables and Attributes-----

#	Variable	Type	Len	Pos	Format
2	absences	Num	8	0	
4	chrclass	Char	1	40	
6	chrdob	Char	8	41	
5	dob	Num	8	16	DATE9.
1	name	Char	8	32	
7	numclass	Num	8	24	
3	quiz	Num	8	8	

Obs	name	absences	quiz	chrclass	dob ❸	chrdob	num class
1	tim lou	1	6.0	3	23OCT1978	10/23/78	3
2	sally	2	10.1	2	02JAN1978	01/02/78	2
3	john	2	8.0	3	22JUN1978	06/22/78	3

In Example 3.9, the PUT() function is used to create a character version of the *dob* variable ❶. The *chrdob* variable is a character variable in the format of MMDDYY8. Notice the difference between the format and display of the *dob* and *chrdob* variables in the output ❸. The INPUT() function is used to convert the character variable *chrclass* to the numeric variable *numclass* ❷. Also note that in both PUT() and INPUT() functions, the format and informat are numeric.

3.7 Processing and Executing SAS Statements Iteratively Using DO Loops

Programming time can be saved by using DO loops for performing repetitive calculations and generating repetitive SAS statements. The alternative to using DO loops is to specify explicitly each SAS statement to execute. **The DO loop controls the number of, and conditions to, execute SAS statements without manually writing or repeating each SAS statement.** It also reduces code by identifying patterns in the repeated code to a standardized block of SAS statements.

3.7.1 DO Loops

Syntax:
DO index-variable= start to end BY count;   */* Top evaluation */*
END;

DO index-variable= specification-1...specification-n; */* Top
                                                                                evaluation */*

END;

DO UNTIL (expression);   */* Bottom evaluation */*
END;

DO WHILE (expression);   */* Top evaluation */*
END;

There are three types of DO loops that will be reviewed in this section. The first type is the iterative DO statement. The specifications in the iterative DO loop can be a list of values, constants, or starting and ending values. If the BY clause is not specified, then the default increment value is 1. The iterative DO statement executes SAS statements between the DO and the END statements iteratively until the number specified is reached. SAS creates the index variable and increments it according to the specifications listed. Thus, this index variable must be dropped from the data set if it is not needed. In addition, SAS processes DO loops as if the code between the DO and END statements is repeated by the number of specifications listed. The code is repeated in the same sequence as was originally specified.

The next two types of DO statements are the DO UNTIL and the DO WHILE statements. Both of these types are similar in that the DO loop will execute until a condition is met. The difference is that with the **DO UNTIL statement, the condition will be evaluated at the bottom of the loop** while the **DO WHILE statement will evaluate the condition at the top of the loop.** Thus, if a condition is false, the DO UNTIL statement will execute all the statements within the loop at least once as compared to the DO WHILE statement which will not execute at all. Note that with the DO UNTIL and DO WHILE statements, you will need to include a statement to increment the index variable. Remember that if any of the required clauses are missing from the DO statements, SAS will issue a NOTE indicating unclosed DO blocks. Note that DO loops can run within other DO loops.

Example 3.10   Processing data with iterative DO loops.

```
data tests;  ❶
   input student @;     /* Comparable SAS statements without
                ❷          DO loop   */
     do score=1 to 4;   /* score=1;  input test @; output;  */
       input test @;    /* score=2;  input test @; output;  */
       output;          /* score=3;  input test @; output;  */
     end;               /* score=4;  input test @; output;  */
   cards;
   1   7 10   10   8
   2   6  7    7   8
   3 10 10    9  10
run;

proc print data=tests;
run;
```

Output

Obs	student	❸ score	test
1	1	1	7
2	1	2	10
3	1	3	10
4	1	4	8
5	2	1	6
6	2	2	7

7	2	3	7
8	2	4	8
9	3	1	10
10	3	2	10
11	3	3	9
12	3	4	10

In Example 3.10, the iterative DO loop will evaluate the statements inside the DO loop for the number of times specified by the index variable *score*. In this example the first INPUT statement will read the student number. The @ sign at the end of the line instructs SAS to keep reading from the same line of data. ❶ The DO loop will then execute four times, processing the statements within the loop. ❷ When the loop is started, the second INPUT statement will execute, reading the next value. The OUTPUT statement then instructs SAS to write the data to the data set. The DO loop will continue reading the next values of raw data and outputting them to the data set. Note that for each student, SAS assigns the *score* variable values from 1 to 4 for each test value. ❸ The *score* variable is the index variable and saved with the data set.

Example 3.11   Processing data using DO UNTIL and DO WHILE loops.

```
data test;              /* Do until - bottom evaluation */
   n=1;                 /* Comparable SAS statements without
            ❶                DO Loop */
   do until (n>=5);     /* student=1; output; n=n+1; */
      student=n;        /* student=2; output; n=n+1; */
      output;           /* student=3; output; n=n+1; */
      n+1; ❸            /* student=4; output; n=n+1; */
   end;
run;


proc print data=test;
run;


data test;              /* Do while - top evaluation */
   n=1;                 /* Comparable SAS statements without
            ❷                DO Loop */
   do while (n<5);      /* student=1; output; n=n+1; */
      student=n;        /* student=2; output; n=n+1; */
      output;           /* student=3; output; n=n+1; */
```

```
        n+1;        ❸              /* student=4; output; n=n+1; */
      end;
run;

proc print data=test;
run;
```

Output

```
Obs      n          student

 1       1            1
 2       2            2
 3       3            3
 4       4            4

Obs      n          student

 1       1            1
 2       2            2
 3       3            3
 4       4            4
```

In Example 3.11, both the DO UNTIL and the DO WHILE statements give the same result but with different methodology. The expressions needed for the loops are different since the DO UNTIL will evaluate the expression at the bottom of the loop and will execute until the expression is true. Thus, using the expression (n>=5) stops SAS when *n* reaches the value of 5 at the bottom of the DO loop. ❶ The DO WHILE loop will evaluate at the top of the loop and will execute while the expression is true. Using the expression (n<5) stops SAS when *n* reaches the value of 5 at the top of the DO loop. ❷ Note that because an index variable was not used, SAS does not create an index variable. The variable *n* is created before the DO loop and is used in the expression evaluated by the DO loop. In addition, the variable *n* needs to be incrimented in the DO loop.❸

Example 3.12 Processing data using DO UNTIL loop with same condition as DO WHILE loop.

```
data test;                    /* Do until - bottom evaluation */
   n=1;                       /* Comparable SAS statements without
                                 DO Loop */
```

```
do until (n< 5);
   student=n;            /* student=1; output; n=n+1; */
   output;
   n+1;
end;
run;

proc print data=test;
run;
```

Output

```
Obs      n      student

 1       1        1
```

In Example 3.12, the condition on the DO UNTIL statement is the same as the DO WHILE statement in Example 3.11. ❶ Since the condition n < 5 is already true, SAS executes the statements within the DO UNTIL loop only once. Because the DO UNTIL loop performs bottom evaluation, the comparable SAS statements STUDENT = 1; OUTPUT; and N+1; are executed and then SAS exits the DO loop. Thus only one record is created with the value 1 in each variable.

3.8 Processing Data Using SAS Arrays

An array is a collection of SAS variables or initialized values that are grouped under a single name for ease of reference for the duration of the DATA step. An array is a compile time statement and is referenced by the array name and index value. Arrays are useful for many things such as performing repetitive calculations or creating many variables with the same attributes. Arrays can also be used to read data, compare variables, and transpose data. Using arrays offers alternatives to specifying variable names directly. In addition, arrays create variables if they do not already exist in the PDV.

3.8.1 Using SAS Arrays

Syntax:
Explicit Array:
ARRAY array-name {subscript} <$> <array-elements> (initial values);

Implicit Array:
ARRAY array-name {index-variable} <$> <array-elements> (initial values);

There are two types of arrays that can be used, explicit and implicit. Explicit arrays are those in which the number of elements in the array is explicitly specified, while implicit arrays do not specify the number of elements but have an index variable. This index variable is used to reference implicit array elements. **The ARRAY statement must contain three items: the array name, a subscript that indicates the number of elements in the array, and a listing of the array elements.** If the list of array elements is not specified, then, by default, SAS uses the array name as the root name with a numeric suffix for each element. All array elements must be either numeric or character, and the different element types cannot be used by the same array name. In addition, the array may contain the initial values for the elements.

The array name is used to identify the array and is a name that you can choose. It is best not to give an array the same name as a variable. Following the name in explicit arrays is the subscript, which is an asterisk, a number or a range of numbers if the array is a multidimensional array. In an implicit array, the index variable is listed. If the elements in the array are characters and have not been previously defined, then a "$" comes before the array elements. **Note that SAS will issue an error message if you try to reference an array element that has not been defined.** Make sure you do not accidentally go beyond the array elements when using the DO loop to process arrays. See the SAS paper *Hurray for Arrays!* for more information on SAS arrays.

Example 3.13   Processing variables with arrays.

```
data class3;
   input name $8. @10 absences 2. @15 quiz 4.1;
cards;
tim lou       1          6.0
sally         2         10.1
john          2          8.0
;
run;


data class_array;
  set class3;


* Creates character variables var1, var2, and var3 and
assign to the value of I;
* The default length of 8 is applied to thesevariables;
* array name vars;
array vars(3) $ var1-var3; ❶
```

```
do I = 1 to 3;  /* index value */
  vars(i) = I;          /* single statement */
end;
/* Each array element references a variable.   Similar to
three separate statements */
/* Ex. var(1) is var1; var1=1; var2=2; var3=3; */ ❷
* implicit array with variable names absences and quiz;
array factors(*) absences quiz; /*existing data set
              ❸                      variables are
                                     referenced*/
do I = 1 to dim(factors);         /* resolves to 2 for 2
                                     elements in array */
  Factors(i) = factors(i)*10;     /* single statement */
end;
/* Similar to two separate statements */
/* absences = absences*10; quiz = quiz*10; */ ❹

* Create variables test85, test86, test87;
array test{85:87}test85 - test87; /* Single statement */ ❺

/* Similar to three separate statements */
/* test85 = .; test86 = .; test87 = .; */ ❻

run;

proc print data=class_array;
run;
```

Output

Obs	name	absences	quiz	var1	var2	var3	I	test85	test86	test87
1	tim lou	10	60	1	2	3	3	.	.	.
2	sally	20	101	1	2	3	3	.	.	.
3	john	20	80	1	2	3	3	.	.	.

In Example 3.13, there are three different ARRAY statements that demonstrate the different ways that an ARRAY statement can be used. In the first ARRAY statement, the array *vars* is created with 3 elements, which are characters and are named *var1*, *var2*, and *var3*. ❶ ARRAY statements are usually used with DO loops in order to access the elements. In the DO loop following the ARRAY statement, all of the variables are initialized to the index value. As shown in this example, this process would replace three different SAS statements ❷. Although in this example you are not saving much typing by using the array, arrays can be a good time saver if there are a lot of variables.

The second ARRAY statement uses the asterisk in place of a number for the subscript ❸. If the asterisk is used, then SAS will determine the number of elements in the array by counting the elements in the array. In this example, there are two elements. In the DO loop, there is a new function that is used, the DIM() function. The DIM() function resolves to the number of elements that are in the array that is specified. In this example, the array *factors* have two elements, so the DIM() function will resolve to two. Without the DO loop and factors array, you would have to specify the two SAS statements to get the same results.❹

The last ARRAY statement creates the array *test*, which contains three variables (*test85, test86, test87*) ❺. In this ARRAY statement, a range is used to specify the number of elements. The lower and upper ranges are specified and separated by a colon. Since nothing else is done with these variables, they remain initialized with a value of missing. ❻

Example 3.14   Creating temporary arrays and initialized values.

```
data class3;
   input name $8. @10 absences 2. @15 quiz 4.1;
cards;
tim lou   1     6.0
sally     2    10.1
john      2     8.0
;
run;


data class_temp;
  set class3;

* numeric array with initialized values;
array goal{4} g1 - g4   (10, 15, 15, 10); /* Single
statement */ ❶
/* Four separate statements */
/* g1 = 10; g2=15; g3=15; g4=10; */

* character array with initialized values;
* does not create variables ;
array namval{3} $7 _temporary_ ('sally', 'tim lou',
'john'); ❷
do i=1 to 3;
 if trim(left(name)) = trim(left(namval(i))) then catorder
= I; /* single statement */
end;
```

```
/* Three separate statements */
/* if trim(left(name)) = trim(left('sally')) then catorder
= 1; */
/* if trim(left(name)) = trim(left('tim lou')) then
catorder = 2; */
/* if trim(left(name)) = trim(left('john')) then catorder
= 3; */


array transdt(3);   /* Notice that the array elements are
                       NOT explicitly specified. */
/* This is equivalent to the following statement: */ ❸
/* array transdt(3) transdt1 transdt2 transdt3;*/

run;


proc print data=class_temp;
run;
```

Output

Obs	name	absences	quiz	g1	g2	g3	g4
1	tim lou	1	6.0	10	15	15	10
2	sally	2	10.1	10	15	15	10
3	john	2	8.0	10	15	15	10

i	catorder	transdt1	transdt2	transdt3
4	2	.	.	.
4	1	.	.	.
4	3	.	.	.

In Example 3.14, there are also three different ARRAY statements. The first ARRAY statement creates the array *goal*, which has four elements that are initialized to specified values ❶. This single statement would take the place of four different SAS statements. The second ARRAY statement creates a temporary array because of the keyword _TEMPORARY_ ❷. With a temporary array, no variables are created. In the output you can see that the *catorder* variable is created but there are no variables from the ARRAY statement. The benefit of this approach is that array element values can be processed within a DO loop without creating variables to store these values. The last array shows how to create an array where no elements are explicitly specified ❸. The *transdt* aray is created with three elements. Since no array elements were specified, SAS will take the array name and append the element number to create the element names *transdt1*, *transdt2*, and *transdt3*.

3.9 Sorting Observations in a SAS Data Set

3.9.1 The SORT Procedure

Syntax:
PROC SORT DATA=input data set <options> OUT=output data set;
  BY <variable1 ... variablen>;
RUN;

The SORT procedure is used to rearrange observations in a data set in ascending or descending order of the variables specified in the BY statement. By default, SAS will sort the observations in ascending order, but the keyword DESCENDING can be added before the variable's name in the BY statement to sort in descending order. The SORT procedure can sort by one or by multiple variables. If the OUT= option is specified, then the changes will be saved to another data set, or if the option is not used, then the changes will be made permanently to the input data set. The SORT procedure does not generate any output or change any of the data values but only changes the order of the records. If the variable that is being sorted contains any missing values, they will be treated as the lowest values.

If the DATA = option is missing in the SORT procedure, the most recently created data set will be the one sorted. The SORT procedure can also remove duplicate values, if they exist, with the NODUPKEY option. The NODUPKEY option deletes observations with duplicate BY variables in the data set. This works well if you only want to keep the BY variables in the sorted data set. However, the NODUPKEY option should be used with caution. If the records are not true duplicates, in that the variables not specified in the BY statement are different, you will have no control over which observations get removed with the NODUPKEY option.

As an alternative to the NODUPKEY option, SAS offers the NODUPLI-CATES option. With this option, SAS removes adjacent duplicate observations as it writes to the output data set. In order for this approach to be effective, all the variables in the data set must be specified in the BY statement. Note that NODUP and NODUPRECS are alias keywords for the NODUPLICATES option.

To specify all variables in the data set without listing them, you can use the _ALL_ keyword in the BY statement. If the _ALL_ keyword is used, then the data set will be sorted by all of the variables in the data set. See the SAS paper *The Problem with Noduplicates* for more information on the NODUPKEY and NODUPLICATES options. One of the very useful features in Version 9.1 is the ability to save duplicate records in a data set with the DUPOUT= option. See Section 6.3 of Chapter 6 for more information on the DUPOUT= option.

Example 3.15   Using the SORT procedure to sort data sets.

```
data tests;
   input name $ class $ test_score ;
   cards;
   Tim math 9
   Tim history 7
   Tim science 7
   Sally math 10
   Sally science 7
   Sally history 5
   John math 8
   John history 7
   John   science 6
;
run;

proc sort data=tests out=tests_sort; ❶
   by name descending class; ❷
run;

proc print data=tests_sort;
run;
```

Output

Obs	name	class	score
1	John	science	6
2	John	math	8
3	John	history	7
4	Sally	science	7
5	Sally	math	10
6	Sally	history	5
7	Tim	science	7
8	Tim	math	9
9	Tim	history	7

Example 3.15 shows how to sort a data set and send the output to a new data set. The OUT= option on the PROC SORT statement is used to specify an output data set to send the results to. ❶ In the BY statement, the *name* and *class* variables are specified. SAS will first sort the data by the *name* variable and then by the *class* variable. ❷ The *name* variable will be

sorted in ascending order by default, while the *class* variable will be sorted in descending order since the keyword DESCENDING was specified.

3.10 Using SAS Utility Procedures to Investigate SAS Data Libraries

3.10.1 The APPEND Procedure

Syntax:
PROC APPEND NEW=<data set b> BASE=<data set a> <FORCE>;
RUN;
 or
PROC APPEND DATA=<data set b> OUT=<data set a> <FORCE>;
RUN;

The APPEND procedure will concatenate data sets together. The records from the NEW or DATA data set will be added to the end of the records from the BASE or OUT data set records. For variables that are common between the two data sets, make sure the attributes are the same to prevent SAS from generating a WARNING message. **If there are different variables in the data sets, then the FORCE option is required to include all variables.** All data values from both data sets will be included unless otherwise specified. A WHERE statement can be used with the APPEND procedure to restrict the observations included. The APPEND procedure is similar to the SET statement in a DATA step and is ideal for concatenating similar data sets. **The APPEND procedure is better since it does not have to read all of the observations in the original data set before adding the new observations like the SET statement does in the DATA step.** This is important if you are dealing with large data sets since the time to run a program would be longer if a SET statement is used versus the APPEND procedure.

Example 3.16   Using the APPEND procedure to add data sets.

```
data A; ❶
  input patno source $;
  cards;
1 A
3 A
;
run;

data B; ❷
  input patno source $;
  cards;
2 B
4 B
;
run;

❸proc append data=b out=a; /* Comparable DATA step
run;                         /* Data a; set a b; run; */

proc print data=a;
run;
```

Output

Obs	patno	source ❹
1	1	A
2	3	A
3	2	B
4	4	B

In Example 3.16, *a* and *b* data sets are similar in structure ❶❷. Because these data sets have common variables, using the APPEND procedure works well to combine these data sets. Records from the *b* data set are added to the end of the *a* data set and saved to the *a* data set ❸. The order of the records can be seen from the PRINT procedure ❹.

3.10.2 *The COPY Procedure*

Syntax:

LIBNAME input library;
LIBNAME output library;

PROC COPY IN=input library OUT=output library;
  MEMTYPE=(memtype list);
RUN;

The COPY procedure is used to copy the contents of one library to another library. **The COPY procedure is useful if you need to transport SAS data sets from one computer to another.** For the COPY procedure to function properly, both the input and output libraries need to be assigned with a LIBNAME statement. When the libraries have been assigned, the COPY procedure can be called. After the procedure name, the input and output libraries are specified with the IN= and OUT= options. The MEM-TYPE statement is then used to specify the file type you want to copy. Usually DATA is specified as the memtype to copy only the SAS data sets from the input library.

Example 3.17   Using the COPY procedure to copy multiple data sets.
```
libname inx   'c:\students';
libname outx 'c:\grades';
```

```
proc copy in=inx out=outx
   memtype=data;
run;
```

Example 3.17 shows how to copy data sets from the students directory to the grades directory. The libraries that are created with the LIBNAME statement are then used with the IN= and OUT= options to specify the input and output library. Specifying MEMTYPE to equal DATA causes SAS to copy only the data sets.

3.10.3 PROC DATASETS

Syntax:
PROC DATASETS LIBRARY=library name;
AGE current name;
APPEND BASE=data set DATA=data set;
CONTENTS DATA=member;
COPY OUT=libref;
DELETE member list;
EXCHANGE;

MODIFY member name;
  RENAME old name=new name;
  LABEL variable='label';
  INFORMAT variable informat;
  FORMAT variable format;

REPAIR member name;

SAVE member list;

QUIT;

The DATASETS procedure is used to manage data sets within SAS libraries. With the DATASETS procedure, you can copy, rename, or delete SAS data sets. In addition, you can also change variable names and attributes within an SAS data set. Note that the LABEL and FORMAT statements must use the new variable name if these statements follow the RENAME statement. If these statements are placed before the RENAME statement, then use the original variable names. The APPEND statement is similar to the APPEND procedure.

Example 3.18   Using the DATASETS procedure to manage
                 data sets.

```
library compat 'c:\mydata';
proc datasets library=compat;

change old=new;            /* rename file */
exchange member1=member2;
                     /*swaps the names of a pair of members */

age today day1 day2;      /* use to make backup versions of
                             data sets */
delete temp1;             /* delete files */
copy out=;                /* similar to proc copy */

append data=data1 out=data2 force; /* similar to proc append */
contents data=ae;            /* similar to proc contents */

modify ae                 /* data set name */
     rename subject=pt;   /* variable name change */
     label pt='Patient';  /* label change */
     informat pt 5.;      /* informat change */
     format pt 5.;        /* format change */
quit;
```

In Example 3.18, you can see some of the options that are available with the DATASETS procedure. The CHANGE statement option is used to rename a data set. In this case, the old data set is renamed to new. The EXCHANGE statement will exchange the names of two different members. The AGE statement is used to rename a group of members. In this example, the data set *today* is renamed to *day1* and the data set *day1* is renamed to *day2*. The data set that was named *day2* is deleted. This process is helpful for creating backup data sets. The DELETE statement is used to delete members; in this case *temp1* is deleted. The COPY, APPEND, and CONTENTS statements are similar to the COPY, APPEND, and CONTENTS procedures. The MODIFY statement is used to change the attributes of a variable in the specified data set. Each data set that you want to change must have its own MODIFY statement. In this example, the variable *subject* is renamed to *pt* and the label, format, and informat are changed for the *pt* variable.

As an alternative to specifying the data sets manually in order to back up using the AGE statement, you should take advantage of SAS's feature for backing up data with generation data sets and tracking data updates with audit trails. See these Version 8.2 features in Chapter 6, Section 6.2.

Chapter 3.   Managing Data—Chapter Summary

User-Defined Formats

```
proc format library = myformat;
  value $ namefmt   'sally' = 'susan'
                    other  = 'other';

  value classfmt   2 = 'freshman'
                   3 = 'junior'
             other = 'other';

  value qsgrp     0 - 9 = '0 to 9'
               10 - 29 = '10 to 29';
quit;

data test;
  set quiz;

  format name $namefmt. class classfmt. quiz qsgrp. dob
date9.;
run;
```

SAS Statements within DATA Step

DATA Step Task	Example

Assigning variable attributes	`label name = 'Student Name';`
	`format name $8.;`
	`length name $8;`
	`rename gender = sex;`
	`attrib name label='Student`
	`Name' format=$8.`
	`length=$8;`

Sorting observations	`by name; if first.name; /*may`
	` require proc sort in`
	` advance */`

Assigning variable assignments	`retain count 0; count=`
	` sum(count + 1);`
	`age = today() - dob;`
	`state = 'ca';`
	` format age 3. dob mmddyy10.;`

SAS Statements within DATA Step (*continued*)

DATA Step Task	Example

Applying conditional execution	`if absences=2 then class=4;`
	`select (quiz);`
	`when(10.1) grade='A';`
	`when(6.0) grade='B';end;`

Applying functions	`livepnum = index(' I live`
	`here ', 'live');`
	`firsttkn=scan(' I live here`
	`', 1);`

Applying DO loops	`do score=1 to 4 by 1;` `Input test @; output; end;`
Applying arrays	`array var(3) $ var1 var2` `var3;`

SAS Functions

Type	Example
Character	`state=substr('new york,` `ny',5,4);`
Numeric	`xmin= min(x, y, z);`
Date	`dtdiff=intck('week',` `startdt, stopdt);`
Numeric to character	`chrdob =` `put(dob, mmddyy8.);`
Character to numeric	`numclass =` `input(chrclass, 4.);`

DO Loop

Type	Example	Description

DO TO	`do score=1 to 4 by 1;`
	`Input test @;`
	`output; end;`
		Top evaluation.

DO UNTIL	`do until(n >= 5);`
	`n + 1;`
	`end;`
		Bottom evaluation. Specify expression to exit before repeating code.

DO WHILE	`do while(n < 5);`
	`n + 1;`
	`end;`
		Top evaluation. Specify expression to exit after last run.

Arrays

Example	Description
Array var(3) $ var1-var3;	`var1; var2; var3;`
Array goal{4} g1 –g4 (10, 15, 15, 10)	`g1=10; g2=15; g3=15; g4=10;`

SAS Utility Procedures

SAS Procedure	Example	Description

APPEND	proc append data=B
	  out=A;
	run;
	proc append new=B
	  base=A;
	run;
		Adds data set *b* records to end of data set *a*

COPY	libname inx '';
	libname outx '';
	proc copy in=inx
	  out=outx;
	  memtype=data;
	run;
		Selects only data sets to copy from in= to out=library

DATASETS	proc datasets
	  library=compat;
	  change old=new;
	  exchange
	    member=member;
	  age members;
	  delete members;
	  copy out=;
	  append data= out=
	    force;
	  contents data=;
	  modify member
	    (dataset)
	  rename old=new;
	  label variable='string';
	  informat variables
	    informat;
	  format variable
	    format;
	  run;
		Renames file
		Swaps the names of a pair of members
		Used to make backups
		Deletes files
		Similar to proc copy
		Similar to proc append
		Similar to proc contents
		
		Variable name change
		Label change
		Informat change
		
		Format change

Chapter 3.   Managing Data—Chapter Questions

Question 1: How can you ensure a zero value is returned if all variables have missing values in the following statement: x=SUM(a, b, c); ?

Question 2: What are the names of the sales variables in the following ARRAY statement and DATA step that will be included in the *tsales* variable calculation?

```
data x;
   set y;
   array profit{85:90} sales85-sales90;
   do year=86 to 89;
      tsales=tsales+profit{year};
   end;
 run;
```

Question 3: Is a period required in all informats and formats? If so, why?

Question 4: In general, if an SAS procedure does not have the data set name specified, then what data set is used by the SAS procedure?

Question 5: What is one SAS function that you can use to test for the character variable named LNAME containing a numeric value?

Question 6: Is it possible to use the RETAIN statement to preserve a character string in a character variable?

Question 7: Which is the only SAS function that can be used on the left side of an assignment statement to replace data values?

Question 8: Is it possible to assign multiple labels using a single LABEL statement?

Question 9: Can an array contain both character and numeric data values?

Question 10: When using an ARRAY statement, what are the array element names if the list of array elements is not specified?

Question 11: Of the two types of DO loops, DO UNTIL and DO WHILE, which DO loop evaluates the expression at the top of the DO loop?

Question 12: What is one method for testing whether the character variable *name* contains blanks?

Question 13: What function can you use to shift a string one space to the left?

Question 14: Are multiple variables in a FORMAT, LENGTH, or LABEL statement separated by commas or spaces?

Question 15: When using the PUT() function, does the format always have to be a character format?

Question 16: Name at least one character function that returns a numeric value.

Question 17: What is the problem with the following code?

```
data test;
  set a;
  if x = 1 then y =    'NO';
  else if x = 2 then y = 'YES';
run;
```

Question 18: What benefit does the _TEMPORARY_ option in an array statement provide?

Question 19: Determine the length of the variable *x* in the following example.

```
data _null_;
  x = 'toolong'; /* first value */
  attrib x length = $4;
  len = vlength(x);
  put x= len= ;
run;
```

Question 20: What are three keywords available in the VALUE statement of the FORMAT procedure?

Question 21: Is it possible to write an SAS DO loop statement to process the days of the week?

Question 22: What are the types and lengths of variables *x, y, a,* and *b*?

```
data test;
  length x 3. y $10.;
  a = x; b = y;
run;
```

Question 23: What is one method to test whether a character variable *lname* contains a capital letter?

Question 24: What is the maximum length of a variable's label?

Question 25: How long can format labels be?

Question 26: What are the values of the *index* and *capital* variables at the completion of the following DATA step:

```
data test;
 retain capital;
 capital = 1;

 do index = 1 to 10 by 1;
  capital + 5;
 end;
run;
```

Question 27: What is the length of the character string that results from using most SAS character functions, such as SUBSTR( ), LEFT( ), and TRIM( )?

Question 28: What determines the variable type of the result, numeric or character, when using the INPUT( ) function?

Question 29: What attributes can the LENGTH statement define?

Question 30: Which attribute determines whether a variable is character or numeric?

Question 31: When using the RENAME statement to change a variable's name in the new data set, it is possible to reference the new variable's name within the same DATA step?

Question 32: In general, what is the range of the length for numeric variables?

Question 33: Is it possible to use SAS functions to process multiple variables?

Question 34: Identify two things that DO loops can be used for.

Question 35: What SAS statement is required to accumulate totals?

Question 36: What function can be used to confirm that a string contains the letter "r"?

Question 37: How can you override a permanent format?

Question 38: When sorting a data set with the BY statement, what additional variables are temporarily created in the DATA step?

Question 39: What is the purpose of the SCAN( ) function?

Question 40: When using the DATASETS procedure, is it possible to modify variable attributes of several data sets from a single MODIFY statement?

Question 41: What is the difference in applying a LABEL statement in a DATA step as compared to applying it in an SAS procedure?

Question 42: When using the iterative DO loop, is the index variable a temporary or permanent variable?

Question 43: When specifying combined SAS functions, such as SUBSTR(SCAN( )), which function is applied first?

Question 44: If there is no LENGTH or ATTRIB statement or any reference to variables within a program, how does SAS determine the variable's length?

Question 45: What is the final variable type of the *myword* variable in data sets *testa* and *testb?*

```
data testa;            data testb;
 myword = 'hello';      myword = 1;
 myword = 1;            myword = 'hello';
run;                   run;
```

Question 46: Within a DATA step, is it possible to have a compile-time statement within a conditional executable statement?

Question 47: In general, when using the DO UNTIL or DO WHILE statement, what additional statement is required that is not required with the iterative DO statement?

Chapter 4

Generating Reports

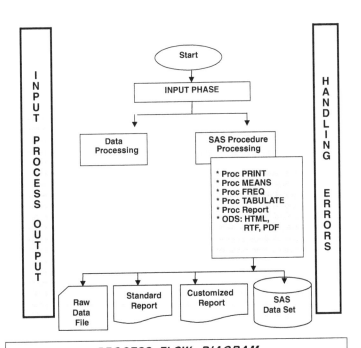

4.1   Introduction

Chapter 4 introduces some of the SAS procedures and options that can be used to produce different types of reports. If you just want to list your data, the PRINT procedure can be used. For summary reports, either the MEANS or the FREQ procedure can be used. The REPORT and TABULATE procedures offer options that allow more control over the look of your output. In general, most SAS procedures, by default, will include missing data if it exists. The MEANS procedure, however, by default, excludes missing values. The last part of this chapter shows how to create HTML, RTF, and PDF reports with ODS. Table 4.1 lists the SAS procedures that will be reviewed in this chapter. The important statements and options of these SAS procedures are reviewed. It is not within the scope of this book to show all of the options available in these SAS procedures. In addition, it is not within the scope of this book to show how to create custom reports using the DATA _NULL_ statement.

4.2   Generating Reports Using the PRINT Procedure

Syntax:
PROC PRINT <options>;
  VAR <variables>;
  ID <variables>;
  BY <variables>;
    PAGEBY <variables>;
    SUMBY <variables>;
  SUM<variables>;
RUN;

The PRINT procedure will be one of the SAS procedures that you use the most. In fact, it was first introduced in Chapter 1. Simply, the PRINT

Table 4.1   Output Procedures

SAS Procedure	Purpose
PRINT	Basic listing of data.
MEANS	Summary statistics of continuous data: N, SUM, MEAN, etc.
FREQ	Cross-tabulation tables with statistics: one-way, *n*-way of categorical variables.
TABULATE	Reports with summary statistics: N, SUM, MEAN, etc.
REPORT	Combines features of PRINT, MEANS, and TABULATE.

procedure prints all the observations in an SAS data set, using all of the variables. There are options and statements that can be added to restrict which variables or observations are printed, order the variables, or perform simple sums. Formats and labels can be temporarily applied to variables to change the display of the column headers and data values. In addition, a WHERE clause may be used to subset the data set.

If only the first line of the PRINT procedure is specified without any options, all of the variables and observations from the most recently created data set will be printed. If the DATA=<data set> option is specified then the PRINT procedure will print the data set that is specified in the option. Using the OBS=n data set option will print the first n observations. In order to select which variables you want printed, the VAR statement can be specified. The PRINT procedure will only print the variables listed in the VAR statement in the same order specified. If the VAR statement is not specified, then SAS displays all variables in the order saved in the data set. The order of variables in a data set can be seen when applying the CONTENTS procedure and viewing the varnum values. The ID statement is used to identify the observations in the output. This is helpful when the list generated contains multiple pages. If you do not use an ID statement, the observation number will identify the observations. The observation numbers can be eliminated by specifying the NOOBS option. By default, the variable name is displayed as a column header. With the LABEL option, SAS displays the variable labels instead. The DOUBLE option will display the list as double spaces. For a consistent alignment across pages, the UNIFORM option should be specified. The WIDTH=MIN option can be added to minimize the space between variables.

The BY statement is used to output a separate analysis for each BY group. If the BY statement is used, the data needs to be sorted in the order of the BY statement with the SORT procedure. If the ID statement and BY statement are both used and specify the same variables, then the report will be grouped by the variables. The PAGEBY statement will cause a new page to be printed when the value of the BY variable changes. In addition, if there are multiple BY variables, a new page will also be printed if any BY variable listed before the PAGEBY variable changes value. The SUMBY statement acts the same way as the PAGEBY statement but instead of printing a new page, the SUMBY statement will print a subtotal. **When the PAGEBY and SUMBY statements are specified, the BY statement must also be specified.** The last statement is the SUM statement. The SUM statement is used to specify which variables are to be totaled. Note that variables in the SUM statement do not need to be specified on the VAR statement since SAS will automatically place them at the end of the report.

Note that when using the BY statement, SAS provides the following keywords to be used in the TITLE statement: #BYLINE, #BYVAR, and #BYVAL.

The #BYLINE keyword in the TITLE statement displays the BY variable name as equal to the value in the title. The #BYVAR keyword in the TITLE statement displays the BY variable name in the title. The #BYVAL keyword in the TITLE statement displays the BY value in the title. The system option NOBYLINE can be used to eliminate the default BYLINE for each group of records.

Example 4.1   Using the PRINT procedure to display data.

```
data tests;
   input name $ class $ test_score;
   cards;
   Tim math 9
   Tim history 7
   Tim science 7
   Sally math 10
   Sally science 7
   Sally history 5
   John math 8
   John history 7
   John  science 6
;
run;

proc sort data=tests;
   by class;
run;

proc print data=tests;
   id name; ❶
   by class; ❷
   sum test_score;
run;
```

Output

```
------------------------------- class=history ---------
                                          test_
                        ❸   name       score
                            Tim          7
                            Sally        5
                            John         7
                            -----      -----
                            class        19 ❹
```

```
-------------------------------- class=math ----------

                              test_
                      name    score

                      Tim       9
                      Sally    10
                      John      8
                      -----    -----
                      class    27

-------------------------------- class=science ---------

                              test_
                      name    score

                      Tim       7
                      Sally     7
                      John      6
                      -----    -----
                      class    20
                               =====
                                66
```

Example 4.1 contains three of the PRINT procedure statements. The ID statement is used to create the identifying variable *name*, instead of using the observation number. ❶ The BY statement groups the variable *class* by the different values of the variable. ❷ Notice that since the *class* variable is used as a BY variable it will not appear with the other variables unless you specify it in a VAR statement. ❸ The SUM statement then summarizes the *test_score* variable for each *class* variable ❹.

4.3 Generating Summary Reports Using the MEANS Procedure

Syntax:
PROC MEANS <options>;
 VAR <variables>;
 BY <variables>;

CLASS <variables>;
OUTPUT <OUT=output data set> <output options>;
RUN;

The MEANS procedure is a powerful SAS procedure used to create summary statistics for your numeric data. Statistics such as n, mean, standard deviation, minimum, and maximum can be created, giving a descriptive view of your data. Using the keywords PROC MEANS instructs SAS to execute the MEANS procedure. There are several options that can be specified after the PROC MEANS keywords. The DATA=<data set> statement can be used to specify the input data set. In addition to the input data set, you can also specify which statistics you want to output. By default, if no statistics are specified, then the MEANS procedure will output the N, MEAN, STANDARD DEVIATION, MIN, and MAX statistics. **Since the MEANS procedure excludes missing values, the statistics produced are based on non-missing values.** The MAXDEC= option can be specified to control the number of decimal places displayed, since, by default, the MEANS procedure displays the full width of each numeric variable.

The VAR statement is used to specify the variables to use as the analysis variables. The order of the variables after the keyword VAR will also be the order that the results are displayed. Only numeric variables can be specified after the VAR statement. If no VAR statement is specified, then the MEANS procedure will output the statistics for all the numeric variables in the data set.

The BY statement performs separate analysis for each level or distinct values of the variables listed. One or more variables can be listed after the BY keyword. The variables must be sorted in the same order specified after the BY keyword. The CLASS statement also performs separate analysis for each variable and combination of variables after the keyword CLASS. Unlike the BY statement, the variables in the CLASS statement do not have to be sorted. In general, the CLASS variables are quantitative variables such as *race* and *gender.*

The OUTPUT statement is used to save the summary statistics to a data set so that it can be used later. The syntax of the OUTPUT statement is the OUTPUT keyword followed by the OUT=<data set> statement. The OUT= <data set> statement is used to specify the name of the output data set that will be created. Following the OUT= statement are the names of the variables that will contain the results. The general form is statistic-keyword(<variable-list>)=<name(s)>. The statistic keyword is the name of the statistic that is being outputted and variable list specifies which of the variables listed in the VAR statement you want to output. New variable names are then specified for each statistic and variable combination.

Although the MEANS procedure is similar to the SUMMARY procedure in syntax and results, the MEANS procedure, by default, produces a report while the SUMMARY procedure, by default,

produces an output data set, unless the PRINT option is specified.
In addition, because of the CLASS statement, the MEANS procedure does
not have to be presorted for BY group processing while the SUMMARY
procedure does require presorting the data set.

**Example 4.2   Using the MEANS procedure to get summarized
              statistics.**

```
data tests;
   input name $ class $ test_score ;
   cards;
   Tim math 9
   Tim history 7
   Tim science 7
   Sally math 10
   Sally science 7
   Sally history 5
   John math 8
   John history 7
   John   science 6
 ;
 run;
```

❶

```
proc means data=tests n mean noprint;
   var test_score;
   class name;
   output out=test_stat n(test_score)=test_score_n
   mean(test_score)=test_score_mean;
run;

proc print data=test_stat;
run;
```

Output

Obs	name	_TYPE_	❷ _FREQ_	test_ score_n	test_ score_ mean
1		0	9	9	7.33333
2	John	1	3	3	7.00000
3	Sally	1	3	3	7.33333
4	Tim	1	3	3	7.66667

Example 4.2 shows the output created by the MEANS procedure. There are a couple of things to note in this example. In the PROC MEANS statement, the NOPRINT option instructs SAS not to print any of the results since we are writing them to a data set. ❶ When using the OUTPUT statement, the output contains all of the variables specified in the statement plus two additional variables, *_TYPE_* and *_FREQ_*. ❷ The *_FREQ_* variable will contain the number of observations used in the statistics generated while the *_TYPE_* variable will depend on the existence of a CLASS or BY statement. Since the CLASS statement was specified, presorting of the test_stat data set was not required. In this example, the value of zero for the _TYPE_ variable signifies the statistics for all nine records.

Example 4.3   Using the MEANS procedure to get summarized statistics by two variables.

```
data drug;
   input gender $ race $ drug $ dose @@;
   cards;
M White Active 5          M White Active 5
F White Placebo 6         M Nonwhite Active 6
M Nonwhite Placebo 4      F White Active 7
M Nonwhite Active 5       M Nonwhite Placebo 7
F White Active 6          F White Active 4
F Nonwhite Placebo 7      F Nonwhite Placebo 3
M Nonwhite Active 4       F Nonwhite Placebo 1
;
run;


proc format;
  value $ genderf 'M'='Male'
                  'F'='Female';
run;


proc means data=drug n mean noprint;
   var dose;
   class drug gender; ❶
   output out=drug_stat n(dose)= dose_n
   mean(dose)=dose_mean;
run;
proc print data=drug_stat;
run;
```

Output

Obs	drug	gender	_TYPE_	dose_ _FREQ_	dose_n	mean
1			0	14	14	5.00000
2		F	1	7	7	4.85714
3		M	1	7	7	5.14286
4	Active		2	8	8	5.25000
5	Placebo		2	6	6	4.66667
6	Active	F	❷ 3	3	3	5.66667
7	Active	M	3	5	5	5.00000
8	Placebo	F	3	4	4	4.25000
9	Placebo	M	3	2	2	5.50000

In Example 4.3, the two variables *drug* and *gender* are crossed to produce the statistics ❶. The MEANS procedure produces statistics for each level — one for *drug* (observations 4 and 5), one for *gender* (observations 2 and 3), and one for both *drug* and *gender* (observations 6–9). The different levels are identified by the _TYPE_ value. The highest level statistic for the *drug* and *gender* combination has a _TYPE_ value equal to 3 ❷. For analysis using multiple BY variables, the NWAY option can be specified to save only the highest combination of classification variables; thus only those records were _TYPE_=3 will be saved to the output data set.

4.4 Generating Summary Reports Using the FREQ Procedure

Syntax:
PROC FREQ <options>;
  TABLES <requests>/<options>;
  BY <variables>;
  OUTPUT <OUT=output data set> <output options>;
RUN;

The FREQ procedure is used to produce one-way to *n*-way cross-tabulation tables. In general, the FREQ procedure is used on quantitative variables such as *race* and *gender* because the number of unique values is limited. Statistics can also be specified, such as Chi-square or Fisher's Exact Test. The FREQ procedure is called by specifying the keywords PROC FREQ followed by any specified options. One of the most common options used is the DATA=<data set> option, which is used to specify the input data set.

You can use the TABLES statement to specify the variables that you want to tabulate. Listing the variables separated by spaces creates a one-way frequency table. Listing the variables with an asterisk separating the variables creates a two-way to *n*-way frequency table. **The order of the two variables determines the layout of the results. The first variable defines the rows and the second variable defines the columns of the tabulation.** If you omit the TABLES statement, the FREQ procedure will create one-way frequency tables for each variable in the data set. In addition, you can specify multiple TABLE statements or list all variables to tabulate on a single TABLE statement. The difference is the need to specify different options or analysis for each table. For example, common analyses specified are EXACT for Fisher's exact test, CHISQ for the chi-square test, and CMH for the Cochran–Mantel–Haenszel statistic. These options are specified after the "/" (slash symbol) on the TABLES statement. By default, variable labels are used as column headers.

The BY and OUTPUT statements are similar to what was described in the MEANS procedure. As before, if a BY statement is used then the data must be sorted by the BY variables. The syntax of the OUTPUT statement is a little different than the syntax used with the MEANS procedure. As with the MEANS procedure, OUT=<data set> is used to specify the output data set. The difference is that with the FREQ procedure, if the OUTPUT statement is specified, then in the TABLES statement you must also list the analyses that you want performed.

Example 4.4   Using the FREQ procedure to summarize data.

```
data tests;
   input sex $ eye_color $ @@;
   cards;
   M Green F Brown M Brown F Green
   F Brown M Green F Brown M Brown
   F Brown M Green F Green F Brown
   F Green F Brown F Green F Green
   M Green M Green M Green F Green
   M Green M Brown M Green M Brown

   ;
run;

proc freq data=tests;
   tables sex*eye_color/exact;
   output out=test_freq exact;
run;                          ❶
```

```
proc freq data=tests;
   tables sex*eye_color/list;
run;

proc print data=test_freq;
run;
```

Output

```
                The FREQ Procedure
             Table of sex by eye_color
         sex              eye_color
        ❷ Frequency
          Percent
          Row  Pct
          Col  Pct       Brown     Green     Total
        ─────────────────────────────────────────────
          F            ❸   6          6          12
                        25.00     25.00      50.00
                        50.00     50.00
                        60.00     42.86
        ─────────────────────────────────────────────
          M                4          8         12
                        16.67     33.33      50.00
                        33.33     66.67
                        40.00     57.14
        ─────────────────────────────────────────────
          Total           10         14        24
                        41.67     58.33     100.00
```

```
        Statistics for Table of sex by eye_color
 Statistic                         DF     Value      Prob
 ─────────────────────────────────────────────────────────

 Chi-Square                         1     0.6857    0.4076
 Likelihood Ratio Chi-Square        1     0.6894    0.4064
 Continuity Adj. Chi-Square         1     0.1714    0.6788
 Mantel-Haenszel Chi-Square         1     0.6571    0.4176
 Phi Coefficient                          0.1690
 Contingency Coefficient                  0.1667
 Cramer's V                               0.1690
```

```
                        Fisher's  Exact  Test
         _____

         Cell  (1,1)  Frequency  (F)        6
         Left-sided  Pr  <=  F              0.8931
         Right-sided  Pr  >=  F             0.3401

         Table  Probability  (P)           0.2332
         Two-sided  Pr  <=  P              0.6802


                    Sample  Size  =  24

                    The  FREQ  Procedure


                         Cumulative   Cumulative
    sex    eye_color  Frequency   Percent     Frequency   Percent
    _____

     F       Brown        6        25.00          6        25.00
  ❹  F       Green        6        25.00         12        50.00
     M       Brown        4        16.67         16        66.67
     M       Green        8        33.33         24       100.00


         Obs     XPL_FISH     XPR_FISH     XP2_FISH
          1      0.89312      0.34009      0.68017
```

In Example 4.4, you can see how the TABLES and OUTPUT statements are used to create the data set *test_ freq*, which contains the output for the Fisher's Exact test. ❶ In order to output the statistics, the statistical test must be specified in both the OUTPUT and TABLES statements. The FREQ procedure produces a table of frequencies. The information within each cell is labeled as follows: cell frequency, cell percentage of total frequency, cell percentage of row frequency, and cell percentage of column frequency ❷. The values within each cell correspond to these labels ❸. The LIST option is useful for displaying the tabulation results as a listing of all the unique combinations of variables ❹.

4.5 Generating Reports Using the TABULATE Procedure

Syntax:
PROC TABULATE <options>;
  CLASS <variables>;
  VAR <variables>;

**TABLE <page-expression, row-expression, column-expression>/
<options>;**
  <KEYLABEL;>
RUN;

The TABULATE procedure computes many of the statistics that are computed with the MEANS and FREQ procedures, plus it offers powerful and flexible reporting features. The TABULATE procedure is called by specifying the PROC TABULATE statement. The PROC TABULATE statement has several options that can be specified to enhance your report. Several of the more common options are the DATA=<data set> option, which specifies the input data set, the FORMAT = option, which formats all of the cells in the tables, and the MISSING option, which displays missing class values in the report.

The CLASS statement specifies the class variables for the table. The CLASS variables are the variables that the TABULATE procedure will use to group the data. The variables specified in the CLASS statement can be numeric or character with a few discrete values that define the classifications. The VAR statement will then identify the variables that will be used as the analysis variables. **The variables specified in the VAR statement must be numeric since they are the variables for which the statistics will be computed.** Table 4.2 lists valid TABLE statement operators.

The TABLE statement describes the table that will be created. Every TABULATE procedure requires at least one TABLE statement. The variables that are listed in the TABLE statement must be listed in either the VAR or the CLASS statement. Following the TABLE statement are one to three dimension expressions. If all three dimension statements are specified, the leftmost expression is the page expression, followed by the row expression, and then the column expression. If only two expressions are specified, the left expression is the row expression and the right expression

Table 4.2 Valid TABLE Statement Operators

Operator	Symbol	Action
Comma	','	Go to a new dimension (column, row, page) of the table.
Space		Concatenate tables side by side or stacking on top.
Asterisk	*	Cross or subgroup for hierarchy structure.
Parenthesis	()	Group of specify order.
Brackets	<>	Specify denominator definitions.
Equal	=	Assign label or format to preceding variable or statistic.

is the column expression. A single expression defines the columns. When specifying two or more expressions, separate the different expressions with a comma.

There are several operators that can be used in each of the expressions depending on what type of table you want. In the TABLE statement each expression is made up of operators and elements. Elements can be the variables that you specified in the VAR or CLASS statement or they can be select statistics. The TABULATE procedure will perform simple statistics such as N, NMISS, MEAN, STD, MIN, MAX, RANGE, SUM, PCTN, and PCTSUM. For class variables, COUNT is the default statistic. For analysis variables, SUM is the default statistic. **Note that while statistics can be displayed in any dimension, all the statistics must be specified in the same dimension.** If the elements are separated by a space operator, then the tables will be concatenated together. This can either be joining tables side by side or stacking them on top of each other. Specifying the elements with the asterisk operator between them will produce a hierarchical table. Note that SAS does not allow you to cross an analysis variable with another analysis variable. In addition, you cannot cross two or more statistics. Another operator is the parenthesis, which can be used to group different elements together. In addition, with parenthesis, a single operator can affect a group of elements.

The last two operators, brackets and the equal sign, are used to supply additional information. The brackets operator is used with statistics such as PCTN and PCTSUM to specify the denominator that will be used in the statistic. The equal sign is used with the FORMAT and LABEL expressions to assign a format or label to a preceding element. The syntax of the statements are F=option for formats and variable_name='label' for labels. Since analysis variables are numeric, only numeric formats are allowed with the F=format modifier. Finally, the special class temporary variable ALL can be used to request totals. Also, the KEYLABEL statement enables you to label statistics and the ALL class variable.

Example 4.5 Using the TABULATE procedure to produce categorical reports.

```
data drug;
   input gender $ race $ drug $ @@;
   cards;
M White Active        M White Active
F White Placebo       M Nonwhite Active
M Nonwhite Placebo    F White Active
M Nonwhite Active     M Nonwhite Placebo
F White Active        F White Active
```

```
F Nonwhite Placebo      F Nonwhite Placebo
M Nonwhite Active       F Nonwhite Placebo

;
run;


proc tabulate data=drug;
  class gender race drug;❶
  table (gender race),❷
        drug*(n pctn<gender race>='%');❸
run;
```

Output

	drug			
	Active		Placebo	
	N	%	N	%
gender				
F	3.00	37.50	4.00	66.67
M	5.00	62.50	2.00	33.33
race				
Nonwhite	3.00	37.50	5.00	83.33
White	5.00	62.50	1.00	16.67

In Example 4.5, you can see some of the syntax of the TABULATE procedure. The CLASS statement ❶ specifies the classification variables that will be used in the analysis. Since all of the variables are character variables you cannot specify any of the variables in a VAR statement. The TABLE statement defines how the table will look. Since only two expressions are defined, the left expression ❷ is the row expression and the right expression ❸ is the column expression. The row expression uses parentheses, which groups the two variables together, but since the space operator is specified the variables will be stacked. In the column expression, the *drug* variable is the first expression and the N and PCTN statistics make up the second expression. When looking at the column expression, you will notice that PCTN uses the brackets to define which variables will make up the denominators used to calculate the percent. The N and PCTN statistics are then enclosed in parentheses to group them together. Since the space operator is used, the two tables will be concatenated together. The asterisk is then used to combine the drug with the N and PCTN statistics, which creates a subgroup for each category. One last thing to note in the PCTN statistic is that the equal sign is used to specify "%" as the label.

Example 4.6 Using the TABULATE procedure to produce statistical reports.

```
data drug;
   input gender $ race $ drug $ dose @@;
   cards;
M White Active 5        M White Active 5
F White Placebo 6       M Nonwhite Active 6
M Nonwhite Placebo 4 F White Active 7
M Nonwhite Active 5     M Nonwhite Placebo 7
F White Active 6        F White Active 4
F Nonwhite Placebo 7 F Nonwhite Placebo 3
M Nonwhite Active 4     F Nonwhite Placebo 1
;
run;

proc tabulate data=drug;
   class drug;❶
   var dose;
   keyword n mean min std max;
   table dose = 'Dose'*(n='N'*f=15. mean='Mean'*f=15.1
std='STD'*f=4.2 min='Min'*f=4.1 max='Max'*f=4.1),
drug=' ' / box=[label='Baseline'] rts=53; ❷
run;
```

Output

Baseline		Active	Placebo
Dose	N	8	6
	Mean	5.3	4.7
	STD	1.04	2.42
	Min	4.0	1.0
	Max	7.0	7.0

In Example 4.6, the TABULATE procedure produces a statistical report by *drug* on the continuous variable *dose* ❶. The syntax of the TABLE statement is very different from that in Example 4.5 ❷. The difference is that statistics are displayed along rows instead of by columns. The BOX= [LABEL= ' '] option after the slash (/) specifies the label displayed in the upper left corner. The RTS= option specifies the size of the row title space. These are the columns used to display the row titles. The TABULATE procedure offers great flexibility in generating complex reports. See the SAS papers *The Power and Simplicity of the Tabulate Procedure* and *The Utter "Simplicity?" or the Tabulate Procedure — The Final Chapter?* for more information.

4.6 Generating Reports Using the REPORT Procedure

Syntax:
PROC REPORT <options>;
  COLUMN <column-specification>;
  DEFINE report-item / <usage> <attribute-list>;
  COMPUTE <variable name>;
  ENDCOMP;
  <BREAK;>
RUN;

The REPORT procedure combines elements of the PRINT, MEANS, and TABULATE procedures. **The main advantage of the REPORT procedure over the TABULATE procedure is the ability to create new variables dynamically using COMPUTE block statements.** The REPORT procedure is invoked with the PROC REPORT statement. The PROC REPORT statement also has several options that can be specified. The most common PROC REPORT statement options are listed in Table 4.3. See the SAS paper *Battle of the Titans: REPORT vs. TABULATE* for more information about the comparison between the REPORT and TABULATE procedures. See also the SAS paper *Fast Track to PROC REPORT* Results.

The COLUMN statement specifies the variables and their order in the report. The syntax of the COLUMN statement is the keyword COLUMN

Table 4.3   Selected PROC REPORT Statement Options

Option	Action
BOX	Outlines the detail lines of the report similar to PROC TABULATE.
CENTER|NOCENTER	Specifies if the output is centered or left-justified.
DATA=	Specifies input data set.
HEADLINE	Underlines all column headers and the spaces between them.
HEADSKIP	Writes a blank line beneath all column headers.
MISSING	Consider missing values as valid values for group, order, or across variables.
NOWINDOWS	Displays a listing of the report in the OUTPUT window.
SPLIT='split character'	Specifies the split character used to break column headers. The default is "/" and space.
WINDOWS	Invokes the procedure in a user-controlled windowing mode.
WRAP	Prints all row values before going to the next line.

followed by the variables that you want in the report. By default, if no COLUMN statement is specified, then the report will contain a column for each variable in the input data set. The columns will appear in the same order as the variables were saved in the data set. Note that variables used for statistics are also specified in the COLUMN statement.

The DEFINE statement has several features. You must specify a separate DEFINE statement for each variable used in the report. The DEFINE statement assigns formats, column headers, widths, justification, and spacing as options after the slash (/). Any format can be specified with the FORMAT= option. Column header is the label in quotes. The width is specified with the WIDTH= option. The justification can be LEFT, CENTER, or RIGHT. Finally, the SPACING= option can be used to change the column spacing. The DEFINE statements can be in any order and do not have to be in the same order as the COLUMN statement. The syntax of the DEFINE statement is the keyword DEFINE followed by the report item. The report item is the variable that you are defining. Note that the report item can also be a new computed variable. Following the report item is the slash (/) and the usage. The usage defines how you are going to use the variable in the report. For each variable, you can only specify one of the five different usages as displayed in Table 4.4.

Note that all analysis variables must be numeric. Also note that when calculating statistics, if there is no group in the report, then the value of the analysis variable is the value of a single observation.

The COMPUTED statement is used to create variables not already in the input data set. This ability to create new variables is unique to the

Table 4.4   DEFINE Statement Usages

Usage	Action
ACROSS	Uses the variable to form column headers. A column is created for each unique value.
ANALYSIS SUM MEAN	Calculates a statistic for all the observations that have a unique combination of values for all group variables. The default is SUM.
COMPUTED	Specified as a new computed variable.
DISPLAY	Displays the variables as they appear in the data set.
GROUP	Consolidates the values of several rows into one row for all observations that have the same value for the group variable.
ORDER	Determines the order of the rows in the report. Prints first obs only. The default order is by data.

REPORT procedure. Almost any DATA step statement used to create numeric or character variables can be specified within the COMPUTED block statements. Within the COMPUTED block statements, the value in these variables is retained from the previous record. In addition, any new variable created can be formatted, labeled, or summarized as a true data set variable. This requires the COMPUTED usage in the DEFINE statement. Note that the position of the computed new variable is important because SAS does not allow any variable to the right side of the computed variables to be included in the calculation.

The BREAK statement controls what happens as values of the ORDER, GROUP, or ACROSS variable change. The BEFORE or AFTER keyword specifies the break location and the variable name specifies the variable used. This specifies when the break is written. It can be a summary line, a page break, or a line space.

If the BY statement is specified, SAS will create a new table for each BY group. In addition, this requires resorting the data set.

Example 4.7   Using the REPORT procedure to produce a listing.

```
data drug;
   input gender $ race $ drug $ dose @@;
   cards;
M White Active 5      M White Active 5
F White Placebo 6     M Nonwhite Active 6
M Nonwhite Placebo 4 F White Active 7
M Nonwhite Active 5   M Nonwhite Placebo 7
F White Active 6      F White Active 4
F Nonwhite Placebo 7 F Nonwhite Placebo 3
M Nonwhite Active 4   F Nonwhite Placebo 1
;
run;

proc format;
  value $ genderf 'M'='Male'
                  'F'='Female';
run;

proc report data=drug headline headskip missing split='*'
nowindows;❶
  columns drug gender race dose;❷
```

```
  define gender / display format=$genderf. width=8 center
'Baseline*Gender'; ❸
   define race / display width=8 center 'Baseline*Race';
   define drug / group width=7 center 'Study*Drug';❹
   define dose / display width=5 'Study*Drug*Dose';
run;
```

Output

Study Drug	Baseline Gender	Baseline Race	Drug ❶ ❷ Dose
❹	❸		
Active	Male	White	5
	Male	White	5
	Male	Nonwhite	6
	Female	White	7
	Male	Nonwhite	5
	Female	White	6
	Female	White	4
	Male	Nonwhite	4
Placebo	Female	White	6
	Male	Nonwhite	4
	Male	Nonwhite	7
	Female	Nonwhite	7
	Female	Nonwhite	3
	Female	Nonwhite	1

Example 4.7 produces a simple report with four columns. The REPORT statement ❶ includes several options that enhance the report output. The HEADLINE and HEADSKIP options add the line and blank row to the column headers. The split character is redefined as "*" and the MISSING option instructs SAS to include missing values as valid values of the variables. The COLUMN statement ❷ lists the variables that are included in the report along with the order of the variables. The DEFINE statements describe the attributes of the columns. The first DEFINE statement ❸ describes the way that *gender* variable will be displayed. The format of $genderf is applied to the *gender* variable along with the column title "Baseline Gender." In addition, the column is centered with a width of 8 characters. The *drug* DEFINE statement describes how the *drug* variable will be displayed ❹. This DEFINE statement is a little different in that the usage is GROUP. Since the statement is specified with the GROUP usage, only the first record of each group is printed in the report.

Example 4.8   Using the REPORT procedure to produce summary tables of continuous data.

```
data drug;
   input gender $ race $ drug $ dose @@;
   cards;
M White Active 5      M White Active 5
F White Placebo 6     M Nonwhite Active 6
M Nonwhite Placebo 4  F White Active 7
M Nonwhite Active 5   M Nonwhite Placebo 7
F White Active 6      F White Active 4
F Nonwhite Placebo 7  F Nonwhite Placebo 3
M Nonwhite Active 4   F Nonwhite Placebo 1
;
run;

proc format;
  value $ genderf 'M'='Male'
                  'F'='Female';
run;

proc report data=drug headline headskip missing split='*'
nowindows;
  columns drug dose;

  define drug / group width=7 center 'Study*Drug';
  define dose / analysis mean f=5.1 'Study*Drug*Dose Mean';❶
run;
```

Output

```
        Study
                        Drug
                Study   Dose
                Drug    Mean
                ───────────────

                Active   5.3
        ❷       Placebo  4.7
```

Example 4.8 shows how to apply the REPORT procedure to create summary tables of continuous data. The ANALYSIS MEAN option in the DEFINE DOSE statement instructs SAS to generate the mean drug dose ❶. Since the GROUP option is specified on the DEFINE DRUG statement, SAS generates mean dose values for each drug. ❷

Example 4.9 Using the REPORT procedure to produce summary tables of categorical data.

```
data drug;
   input gender $ race $ drug $ dose @@;
   cards;
M White Active 5       M White Active 5
F White Placebo 6      M Nonwhite Active 6
M Nonwhite Placebo 4 F White Active 7
M Nonwhite Active 5    M Nonwhite Placebo 7
F White Active 6       F White Active 4
F Nonwhite Placebo 7 F Nonwhite Placebo 3
M Nonwhite Active 4    F Nonwhite Placebo 1
;
run;

proc format;
  value $ genderf 'M'='Male'
                  'F'='Female';
run;

proc report data=drug headline headskip missing split='*'
nowindows;
   columns drug gender; ❶
   define gender / across  format=$genderf. width=8 center
   'Baseline*Gender';
   define drug    / group width=7 center 'Study*Drug';
run;
```

Output

```
                  Baseline
   Study            Gender
   Drug      Female        Male

   Active        3           5 ❷
   Placebo       4           2
```

Example 4.9 shows how to apply the REPORT procedure to create summary tables of categorical data. The ACROSS option in the DEFINE GENDER statement instructs SAS to generate the count for each unique value of the *gender* variable ❶. Since the GROUP option is specified on the DEFINE DRUG statement, SAS generates gender counts for each drug. ❷

4.7 Generating HTML, RTF, and PDF Reports Using ODS Statements

Syntax:
ODS <destination> FILE=<file name> STYLE=<style name>;
  SAS Procedure;
ODS <destination> CLOSE;

ODS HTML PATH = <physical directory> URL=(<option>)
      BODY = <name>
      FRAME = <name>
      CONTENTS = <name>
      PAGE = <name>

      ;
  SAS Procedure;
ODS HTML CLOSE;

With ODS, SAS enables you to create reports easily in various file types known as destinations, such as HTML, RTF, and PDF. In addition, Excel files can also be created through the HTML destination. The results of most any SAS procedures can be saved as any of these file types. In addition, ODS allows you to select a style for your report from any of the SAS-supplied styles or your custom style. It is not within the scope of this book to show how to create custom styles or to show all of the options available when creating output files.

In order to output the data from an SAS procedure to HTML, RTF, or PDF files, you need to add an ODS statement before the SAS procedure to open each file. For example, the syntax of one statement is ODS HTML FILE='<file name>', where the file name in quotes includes the full path where the file will be located. The RTF and PDF statements are similar. ODS allows you to write to all files simultaneously without having to run the SAS procedure three times. After the SAS procedure, the ODS HTML CLOSE statement is specified to close the HTML file. Likewise, the ODS RTF CLOSE and ODS PDF CLOSE statements close the RTF and PDF files respectively. As an alternative, the ODS CLOSE ALL statement can be specified to close all output files automatically. **Always remember to close the output files before accessing them.** Note that by default, SAS writes to the SAS list file unless the ODS LISTING CLOSE statement is specified. In Version 9, there are additional features available for customizing your HTML, RTF, and PDF files. See Chapter 6, Section 6.3 for more information. See the following SAS papers and book for more information about ODS: *Quick Results with the Output Delivery System, Utilizing Clinical SAS Report Templates with ODS, SAS®'s ODS Technology* *for Today's Decision Makers,* and *Using Styles and Templates to Custom-ize SAS ODS Output.*

Example 4.10   Generating HTML, RTF, and PDF reports with ODS.

```
data drug;
   input gender $ race $ drug $ dose @@;
   cards;
M White Active 5      M White Active 5
F White Placebo 6     M Nonwhite Active 6
M Nonwhite Placebo 4  F White Active 7
M Nonwhite Active 5   M Nonwhite Placebo 7
F White Active 6      F White Active 4
F Nonwhite Placebo 7  F Nonwhite Placebo 3
M Nonwhite Active 4   F Nonwhite Placebo 1
;
run;


proc format;
  value $ genderf 'M'='Male'
                  'F'='Female';
run;

ods html file='c:\drug.html'; ❶
ods rtf file='c:\drug.rtf'; ❷
ods pdf file='c:\drug.pdf'; ❸


proc report data=drug headline headskip missing split='*'
nowindows;
  columns drug gender race dose;
  define gender / display format=$genderf. width=8 center
  'Baseline*Gender';
  define race / display width=8 center 'Baseline*Race';
  define drug / group width=7 center 'Study*Drug';
  define dose / display width=5 'Study*Drug*Dose';
run;
ods pdf close;
ods rtf close;
ods html close;
```

Output: HTML File

The SAS System

Study Drug	Baseline Gender	Baseline Race	Study Drug Dose
Active	Male	White	5
	Male	White	5
	Male	Nonwhite	6
	Female	White	7
	Male	Nonwhite	5
	Female	White	6
	Female	White	4
	Male	Nonwhite	4
Placebo	Female	White	6
	Male	Nonwhite	4
	Male	Nonwhite	7
	Female	Nonwhite	7
	Female	Nonwhite	3
	Female	Nonwhite	1

Output File: RTF File

The SAS System

Study Drug	Baseline Gender	Baseline Race	Study Drug Dose
Active	Male	White	5
	Male	White	5
	Male	Nonwhite	6
	Female	White	7
	Male	Nonwhite	5
	Female	White	6
	Female	White	4
	Male	Nonwhite	4
Placebo	Female	White	6
	Male	Nonwhite	4
	Male	Nonwhite	7
	Female	Nonwhite	7
	Female	Nonwhite	3
	Female	Nonwhite	1

Output: PDF File

The SAS System

Study Drug	Baseline Gender	Baseline Race	Study Drug Dose
Active	Male	White	5
	Male	White	5
	Male	Nonwhite	6
	Female	White	7
	Male	Nonwhite	5
	Female	White	6
	Female	White	4
	Male	Nonwhite	4
Placebo	Female	White	6
	Male	Nonwhite	4
	Male	Nonwhite	7
	Female	Nonwhite	7
	Female	Nonwhite	3
	Female	Nonwhite	1

Example 4.10 shows how ODS saves the output from the REPORT procedure in Example 4.7 to an HTML file (drug.html). ❶ The file can then be opened with an internet browser. In addition, you can also save the results to an Excel file. The only change required would be to change the file name extension from ".html" to ".xls" while still specifying the HTML destination. The RTF file (drug.rtf) ❷ and PDF file (drug.pdf) ❸ are also created. Since the STYLE= option was not specified, ODS uses the default style when creating the output files. Note that although this example shows only one SAS procedure, ODS can save the results from multiple SAS procedures in the same output file.

Example 4.11   Generating an HTML file with table of contents using ODS.

```
data drug;
   input gender $ race $ drug $ dose @@;
   cards;
M White Active 5      M White Active 5
F White Placebo 6     M Nonwhite Active 6
M Nonwhite Placebo 4 F White Active 7
M Nonwhite Active 5   M Nonwhite Placebo 7
F White Active 6      F White Active 4
```

```
F Nonwhite Placebo 7 F Nonwhite Placebo 3
M Nonwhite Active 4  F Nonwhite Placebo 1
;
run;

proc format;
 value $ genderf 'M'='Male'
                 'F'='Female';
run;

ods html path = 'c:\' (url=none)
         body='drug_body.html'  ❶
         frame='drug_frame.html'
         contents='drug_contents.html';

proc report data=drug headline headskip missing split='*'
nowindows;
   columns drug gender race dose;
   define gender / display format=$genderf. width=8 center
   'Baseline*Gender';
   define race / display width=8 center 'Baseline*Race';
   define drug / group width=7 center 'Study*Drug';
   define dose / display width=5 'Study*Drug*Dose';
run;

ods html close;
```

Output: HTML File

Study Drug	Baseline Gender	Baseline Race	Study Drug Dose
Active	Male	White	5
	Male	White	5
	Male	Nonwhite	6
	Female	White	7
	Male	Nonwhite	5
	Female	White	6
	Female	White	4
	Male	Nonwhite	4
Placebo	Female	White	6
	Male	Nonwhite	4
	Male	Nonwhite	7
	Female	Nonwhite	7

Example 4.11 shows how ODS automatically creates the link between the HTML files it creates so that you can easily navigate through the table ❶. The PATH= option specifies the location of the files. The (URL= NONE) option preserves relative link references for ease of portability of the HTML files. The BODY= option specifies the main file that is displayed in the center of the screen. The FRAME= option specifies the HTML file to open with the internet browser. The CONTENTS= option specifies the left table of contents HTML file containing the hyperlink to the specific SAS procedure results. All three of these HTML files are automatically created and you need not have any HTML programming knowledge.

4.8 Enhancing Reports with Options and Statements

Syntax:
TITLE1 'text';   /* **up to 10 titles** */

FOOTNOTE1 'text'; /* up to 10 footnotes */

SAS offers several ways to enhance your reports with options and statements. Titles and footnotes add a descriptive label to the report. They are specified with the TITLE and FOOTNOTE keywords followed by the text in single or double quotes. Up to 10 lines of titles and footnotes are allowed. Make sure balanced quotes exist to prevent SAS from generating ERROR messages.

Once titles and footnotes are specified, they remain in effect for all output until changed or cancelled. Redefining a title or footnote cancels any higher-numbered titles. For example, TITLE3; cancels TITLE4. To cancel all titles and footnotes, specify TITLE; and FOOTNOTE; respectively.

In addition to titles and footnotes, system options can be used to control the layout of the report. For example, the LINESIZE= option specifies the maximum length of the output line, the PAGESIZE= option specifies the maximum number of lines per page of output, and the ORIENTATION= option specifies the layout of the report as PORTRAIT or LANDSCAPE. Other system options such as the following turn ON and OFF when specified: CENTER/NOCENTER, DATE/NODATE, NUMBER/NONUMBER, and BYLINE/NOBYLINE. The CENTER option centers the report. The DATE option displays the SAS date on the report and the NUMBER option displays the page number of the report. The BYLINE option turns on the display of the BY variables in the TITLE statement when SAS procedures include the BY statement. Finally, the PAGENO= option can be used to specify the starting page number to display. Table 4.5 shows the various statements that can be specified in SAS procedures to enhance the report with temporary formats and labels

Table 4.5   Standard SAS Procedure Statements

Statement	Description
Procedure options	Specific to SAS procedure.
VAR	List of variables to display or analyze.
BY	Specify BY group processing.
FORMAT	Temporary assign variable formats.
LABEL	Temporary assign variable label.
WHERE	Subset the data set.

as well as performing BY group processing. The following is a list of SAS books that go into greater detail of the options and of these SAS procedures: *SAS Applications Programming: A Gentle Introduction, Mastering the SAS System, Professional SAS Programming Secrets,* and *SAS Procedures Guide.*

Chapter 4.   Generating Reports—Chapter Summary

SAS Procedure	Example	Description

PROC PRINT	`proc print data=test`
	`  label double`
	`noobs`
	`uniform;`
		Options to use label, double space,
		no observation number, uniform listing.

	`var absences quiz` `  dob;`	List and order of variables to display.
	`id name;`	Useful to link records if number of variables spans across multiple rows.
	`by class;`	BY group processing.
	`format dob date9.;`	Format the display of variables.
	`label dob='Birth';`	Apply variable label.
	`sum quiz;`	Provides the sum of QUIZ variable.
	`where absences < 10;` `run;`	Subset list based on condition.

PROC MEANS	`proc means`
	`data=tests n mean`
	`noprint;`
		Specify statistics to display along with noprint option.

	`  var test_score;`	List of variables to analyze.
	`by class;`	BY group analysis.

	`  output`
	`  out=test_stat`
	`n(test_score)=test_`
	`score_n`
		Save results to output data set.
		Rename N statistics to TEST_SCORE_N.

	`mean(test_score)=`
	`test_score_mean;`
	`run;`
		Rename MEAN statistics to TEST_SCORE_MEAN.

PROC FREQ	`proc freq data=test;`
	`tables`
	`  sex*eye_color/`
	`exact;`
	`run;`
		Lists all unique combinations of SEX and EYE_COLOR along with the Fisher's Exact Test.

(*continued*)

PROC TABULATE	`proc tabulate`
	`data=test`
	`missing`
		Option
		Missing includes missing values in table.

	`class name class;`	Category variables: character or numeric.

	`var quiz;`
	`keylabel all='my`
	`total';`
		Analysis variable: numeric.
		Label a statistic.

	`format class` `classfmt.;`	Format applied.

	`table`
	`name all,`
	`class*quiz*sum all`
	`/ rts=8;`
	`run;`
		Table dimensions—
		Row-expression
		Column-expression
		Options are specified after the slash (/)—row title space.

PROC REPORT	`proc report`
	`  data=test`
	`  nowindows nocenter`
	`headline headskip`
	`missing split='*';`
		Options to run in batch mode, CENTER, HEADLINE, HEADSKIP, include missing values and break column header.

	`column name dob` `  class quiz;`	List and order of variables to display.

	`define name/group`
	`center 'Student`
	` Name';`
		Define each variable's attributes:
		Usage—e.g., group/display

	`define dob/display`
	`center`
	` 'Date*of*Birth'`
	`format=date9.;`
		CENTER/NOCENTER
		Label
		Format

	`where absences < 10;` `run;`	Subset condition

Options and Statements	Example
TITLE	title1 'This is the title of the Report';
FOOTNOTE	footnote1 'This is a footnote.';
LABEL	label name = 'Student Name';
FORMATS	format dob date9.;
OPTIONS	options
	ls=80 /* line size - length of the line used in print files. */
	ps=59 /* page size - the number of lines per page in print output. */
	orientation= landscape /* layout also portrait */
	center /* centers all output */
	date /* print dates */
	number /* page number appears in file. */
	pageno= /* specify starting page number */
	byline /* Procedures print BY lines that identify each BY group. Each BY group starts on a separate page. */ ;

ODS OUTPUT	Example
HTML, RTF, PDF	ods html file='c:\drug.html';
	ods rtf file='c:\drug.rtf';
	ods pdf file='c:\drug.pdf';
	proc print data=drug;
	run;
	ods pdf close;
	ods rtf close;
	ods html close;
HTML	ods html path='c:\' url=(none)
	body='drug_body.html'
	frame='drug_frame.html'
	contents='drug_contents.html';
	proc print data=drug;
	run;
	ods html close;

Chapter 4.   Generating Reports—Chapter Questions

Question 1: What is the difference between the MISSING option in the PROC TABULATE statement and the PRINTMISS option in the TABLE statement?

Question 2: What is the default statistic for using the analysis variable in the TABULATE procedure?

Question 3: In the REPORT procedure, what is the difference between GROUP, DISPLAY, and ACROSS?

Question 4: When using ODS to create various file types such as HTML, PDF, and RTF, can SAS simultaneously create these files without needing to rerun the SAS procedures?

Question 5: Is it possible to specify formats in the PRINT procedure?

Question 6: What are the three dimensions when using the TABULATE procedure, and what is their order of priority?

Question 7: When using the REPORT procedure, if the COLUMN statement is not specified, then in which order are the variables displayed?

Question 8: What is one difference between the MEANS procedure and the SUMMARY procedure?

Question 9: Is it possible to save the results from multiple SAS procedures in the same output file when using ODS?

Question 10: What is the difference between the F= format modifier in the PROC TABULATE statement and the F= format modifier in the TABLE statement?

Question 11: What would be the arrangement of variables if the PRINT procedure did not have a VAR statement?

Question 12: What is the default statistic for using the class variable in the TABULATE procedure?

Question 13: How do you get the PRINT procedure to use the variable's label in the output?

Question 14: In the PRINT procedure, what affect does the BY statement have?

Question 15: Is it possible access an HTML file before executing the ODS HTML CLOSE statement?

Question 16: Does the data set need to be presorted before running the MEANS procedure if using the CLASS statement?

Question 17: What is the default title in all output files?

Question 18: When using the TABULATE procedure, is it possible to have statistics on several dimensions?

Question 19: What is the purpose of the COLUMN statement in the REPORT procedure?

Question 20: In general, which statement is used to save the results of SAS procedures to a data set?

Question 21: What affect does the MAXDEC= option have on the MEANS procedure? How are missing values handled?

Question 22: What is the purpose of the DEFINE statement in the REPORT procedure?

Question 23: What is the REPORT procedure syntax to create the following table? The variable names are drug, gender, race, and dose.

Study Drug	Baseline Gender	Baseline Race	Drug Dose
Active	Male	White	5
	Male	White	5
	Male	Nonwhite	6
	Female	White	7
	Male	Nonwhite	5
	Female	White	6
	Female	White	4
	Male	Nonwhite	4
Placebo	Female	White	6
	Male	Nonwhite	4
	Male	Nonwhite	7
	Female	Nonwhite	7
	Female	Nonwhite	3
	Female	Nonwhite	1

Question 24: When using ODS, how do you change the style of your output file?

Question 25: What is the system option to prevent the display of the BY variable name and value in the title when using the BY statement with the PRINT procedure?

Question 26: What is one benefit for specifying separate TABLES statements when using the FREQ procedure?

Question 27: When using the FREQ procedure to create cross-tabulation results, does the order of the variables affect the layout of the results?

Question 28: When creating a new variable using the COMPUTE block statements in the REPORT procedure, are you required to specify COMPUTED in the DEFINE statement of the new variable?

Chapter 5

Handling Errors

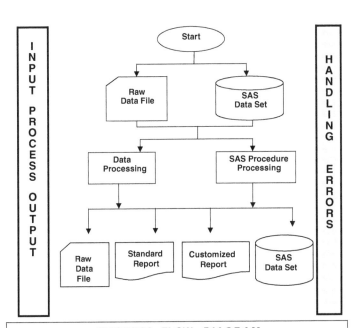

PROCESS FLOW DIAGRAM

PHASE	CHAPTER. DESCRIPTION
INPUT:	1. Accessing Data
	2. Creating Data Structures
PROCESS:	3. Managing and Summarizing Data
OUTPUT:	4. Generating Reports
HANDLING ERROR:	5. Diagnosing and Correcting Errors
V8.2/9.1:	6. Integrity Constraints, Generation DS, Audit Trials

5.1 Introduction

An important but hidden skill required for improved productivity is the ability to diagnose and resolve program problems. By understanding the data and the requirements of the program, you are better able to prevent errors from happening. Chapters 1 through 4 show how to apply SAS syntax to access data, create data structure, manage data, and generate reports. With this knowledge, it becomes easier to understand why you may get errors. Program problems can appear in the SAS log as an ERROR, a WARNING, or a NOTE. The most important rule in debugging your programs is always to check your SAS log for these messages. The only way to know that your program has run correctly is to check the SAS log.

When reviewing your SAS log, you will probably notice ERROR messages first if they exist. ERROR messages need to be corrected first as they will cause the program to abort early. WARNINGs are not as dire as ERROR messages; SAS will print the WARNING message to the SAS log but will continue to run the program. Although your program has finished running, WARNING messages are important because the results could be incorrect due to the problems specified in the WARNING message. NOTEs contain information about the status of your program. They do not necessarily point out the problems in your program but can be used to help debug problems in your program. This chapter focuses on addressing common syntax errors, data related issues, and program debugging techniques. The objective of this chapter is to show a representative sample of the various types of ERRORs, WARNINGs, and NOTEs. It is not within the scope of this book to show all possible ERRORs, WARNINGs, and NOTEs that SAS can generate.

5.2 Recognizing and Correcting Syntax and Non-Syntax Errors

Syntax errors will appear as ERRORs in the SAS log. Non-syntax errors will appear as NOTEs or WARNINGS in the SAS log. Syntax errors result from breaking one of the SAS rules for writing code, such as trying to create a variable starting with a number. Some syntax errors can be difficult to find and may take some investigative work to solve. Syntax errors are compile-time errors, while non-syntax errors are execution-time errors. NOTEs provide basic information about the DATA step or procedure. WARNINGS identify an action SAS took in response to an error triggered in the DATA step. Table 5.1 shows some of the different types of syntax and non-syntax errors and how to resolve them. In the following sections, we will explore each type of syntax and non-syntax error.

Table 5.1   Syntax and Non-Syntax Errors and How to Resolve Them

Syntax Error (Compile Time Error)	*Resolution*
Missing semicolon	Add semicolon
Misspelled/missing keywords or incomplete SAS statement	Correct the keyword spelling or add expected SAS statement
WHERE clause operator requires compatible variables	Ensure both operator and variable types match
At least one W.D. format was too small to be displayed	Change format or increase variable length
Unbalanced quotes	Add missing quote
Non-Syntax Error (Execution Time Error)	*Resolution*
Variable uninitialized and variable not found	Make sure variable exists and is initialized if it is a new variable
MERGE statement has more than one data set with repeats of BY variables	Ensure BY variables uniquely separate records in each data set
Variable type conflict	Ensure common BY variables when merging data sets
BY group processing	Presort data sets before specifying the BY statement in DATA step and SAS procedures

5.2.1 Missing Semicolon

The missing semicolon is one of the most common mistakes made by programmers, even though it is well known that a semicolon is used to end every SAS statement. When you are debugging a program without a semicolon at the end of a SAS statement, messages printed in the SAS log can be difficult to understand because of long miscellaneous and confusing messages.

Example 5.1   Missing semicolon in DATA statement.

```
tests.txt
   Tim math 9 10
   Sally math 10 10
   John math 8 6
----------------------------------------------------

data tests
   infile 'tests.txt';
   input name $ class $ quiz1 quiz2;
run;
```

Log

```
21    data tests
22       infile 'tests.txt';
23       input name $ class $ quiz1 quiz2;
24    run;
```

ERROR: No CARDS or INFILE statement.❶

ERROR: Memtype   field is invalid.

NOTE: The SAS System stopped processing this step because
      of errors.

WARNING: The data set WORK.TESTS may be incomplete. When
         this step was stopped there were 0
         observations and 4 variables.

WARNING: data set WORK.TESTS was not replaced because this
         step was stopped.

WARNING: The data set WORK.INFILE may be incomplete. When
         this step was stopped there were 0
         observations and 4 variables.

**WARNING: data set WORK.INFILE was not replaced because
         this step was stopped.**

NOTE: DATA statement used:
           real time   0.14 seconds
           cpu time   0.09 seconds

The ERROR and WARNING messages produced in Example 5.1 do not really tell what the problem is with the program. The "No CARDS or INFILE statement" message ❶ is odd since there clearly is an INFILE statement in the program. Since the semicolon is missing from the DATA statement, the INFILE statement becomes part of the DATA statement. The word INFILE is then interpreted as another data set name.

To help identify missing semicolons in a program, use the system option DATASTMTCHK=ALLKEYWORDS. With this option, SAS will not accept any keyword as an SAS data set name. By default, you cannot use the words MERGE, RETAIN, SET, or UPDATE as data set names. Below is the same example with the DATASTMTCHK=ALLKEYWORDS option added.

**Example 5.2   Missing semicolon with the system option
                DATASTMTCHK.**

```
options datastmtchk=allkeywords;

data tests
```

```
   infile 'tests.txt';
   input name $ class $ quiz1 quiz2;
run;
```

Log

```
25     options datastmtchk=allkeywords;
26
27     data tests
NOTE: SCL source line.
28        infile 'tests.txt';
           ------
           57
ERROR 57-185: INFILE is not allowed in the DATA statement
when option DATASTMTCHK=ALLKEYWORDS.  Check
              for a missing semicolon in the DATA statement,
or use DATASTMTCHK=NONE. ❶

29        input name $ class $ quiz1 quiz2;
30     run;
```

Example 5.2 produces one ERROR message: "INFILE is not allowed in the DATA statement when option DATASTMTCHK=ALLKEYWORDS option is used." The ERROR message also points you in the correct direction by stating, "Check for a missing semicolon in the DATA statement" ❶.

5.2.2 Misspelled/Missing Keywords or Incomplete SAS Statement

Since SAS expects keywords to process SAS statements, misspelled, missing, or incomplete keywords cause SAS to produce an ERROR message. In addition, SAS also produces an ERROR message if there are any missing required clauses for a block of SAS statements such as the following: IF–THEN–ELSE, SELECT–WHEN, DO loop, etc. For example, SAS gives an ERROR message such as "No matching if-then clause" and a NOTE such as "There were 1 unclosed do blocks" for missing clauses. Similarity, SAS, in general, does not allow DATA step statements to be specified within SAS procedures, unless it is the WHERE statement. Finally, if you try to reference an array element not specified in the ARRAY statement, SAS will issue an "Array subscript out of range" message. Another case of unexpected keyword occurs when a numeric format is applied on a character variable or when a character format is applied on a numeric variable.

Example 5.3   Misspelled MERGE.

```
data C;
  input patno source $ gender $;
  cards;
1 C male
2 C female
3 C male
5 C male
;
run;

data D;
  input patno source $ age;
  cards;
2 D 45
2 D 40
4 D 35
;
run;

proc sort data=C;
  by patno;
run;

proc sort data=D;
  by patno;
run;

data test4;
     merg C D;
     by patno;
run;

proc print data=test4;
run;
```

Output

```
28    data test4;
NOTE: SCL source line.
29    merg C D;
        ----
        1
```

```
WARNING 1-322: Assuming the symbol MERGE was misspelled
as merg. ❶
```

```
30    by patno;
31    run;
```

```
NOTE: There were 4 observations read from the data set
      WORK.C.
NOTE: There were 3 observations read from the data set
      WORK.D.
NOTE: The data set WORK.TEST4 has 6 observations and 4
      variables.
NOTE: DATA statement used:
```

In Example 5.3, SAS will make the correction to the keyword that was misspelled. In this case the keyword MERG was corrected to MERGE. ❶ Even though SAS corrects the spelling, you should go back to your program and make the correction.

5.2.3 WHERE Clause Operator Requires Compatible Variables

The "WHERE clause operator requires compatible variables" message is another ERROR message that can appear in the SAS log. This message is generated by SAS whenever a WHERE statement is used and the operator does not match the variable type. Either the variable is numeric and the operator is character or vice-versa. This ERROR message can be generated in an SAS DATA step or in an SAS procedure where a WHERE clause is used. As discussed in Chapter 2, Section 2.3, all conditional execution statements must have valid expressions in that the variable and operator types must match.

Example 5.4 WHERE clause operator requires compatible variables.

```
data tests;
   input name $ class $ quiz_score ;
   cards;
   Tim math 9
   Tim history 7
   Tim science 7
   Sally math 10
   Sally science 7
```

```
     Sally history 5
     John  math  8
     John  history 7
     John  science 6
   ;
   run;


   proc print data=grade;
      where class=9;
   run;
```

Log (Partial)

```
152   proc print data=grade;
153     where class=9;  ❶
ERROR: Where clause operator requires compatible variables.
      154   run;
```

NOTE: The SAS System stopped processing this step because of
errors.
NOTE: PROCEDURE PRINT used:
           real time   0.06 seconds
           cpu time   0.01 seconds

In Example 5.4, the ERROR message is generated in the PRINT procedure since the variable *class* is a character variable but the operator is
numeric ❶. If you get this message in your SAS log, check the variable
to make sure that you are specifying the correct variable and that the
operator type matches the variable type. In Example 5.4, the *class* variable
is character so the operator needs to be a character value such as "math."
Remember that character data is case-sensitive.

5.2.4 At Least One W.D. Format Was Too Small for the Number to Be Printed

"At least one W.D. format was too small for the number to be printed" is a
NOTE that will appear in your SAS log if you have specified a format that is
too small for the variable. When you have this NOTE in your SAS log, you
are at risk of losing data in your report since SAS will only set aside in memory
the value specified in the format; if the data values are larger than the format,
there may be a loss of data in the report.

Example 5.5 At least one W.D. format was too small for the number to be printed.

```
data tests;
   input name $ class $ quiz1 $  quiz2 $;
   cards;
   Tim math 90 100
   Sally math 100 100
   John math 80 60
;
run;

data total;
   set tests;
   format total 2.; ❶
   total=quiz1 + quiz2;
run;

proc print data=total;
run;
```

Log (Partial)

```
240   proc print data=total;
241   run;
NOTE: There were 3 observations read from the data set
      WORK.TOTAL.
NOTE: At least one W.D format was too small for the number
      to be printed. The decimal may be shifted by
      the "BEST" format. ❷
NOTE: PROCEDURE PRINT used:
      real time              0.01 seconds
      cpu time               0.01 seconds
```

Output

Obs	name	class	quiz1	quiz2	total
S	Tim	math	90	100	** ❸
2	Sally	math	100	100	**
3	John	math	80	60	**

In Example 5.5, the format of 2. is applied to the *total* variable ❶. When the DATA step executes, the values for *total* are created by adding the *quiz1* and *quiz2* variables. When the *quiz* variables are added together the values are larger than two digits, causing the NOTE to be printed in the SAS log. SAS was able to display the variable by changing the format of the total variable to the BEST format ❷. Thus, instead of the actual total value, asterisks are displayed in the output ❸. You should remove this message when you see it in the SAS log by correcting the format, instead of having SAS do it. In this example, the format for the *total* variable should be at least 3. Note that this can also happen with character variables when a format smaller than the variable format is used to display the values. As discussed in Chapter 3, Section 3.2, the length and format for each variable should be large enough to store and display character and numeric data correctly.

5.2.5   Unbalanced Quotes

Unbalanced quotes usually occur in titles and footnotes where the title is enclosed in quotes but one of the quotes is missing. This can also occur in conditional statements where a character string is specified in quotes.

Example 5.6   Unbalanced quotes.

```
title1 'Print of combined data;

data C;
  input patno source $ gender $;
  cards;
1  C  male
2  C  female
3  C  male
5  C  male
;
run;

data D;
  input patno source $ age;
  cards;
2  D  45
2  D  40
4  D  35
;
run;

data test3;
   merge C D;
run;
```

```
proc print data=test3;
run;
```

Log

```
title1 'Print of combined data; ❶

data C;
  input patno source $ gender $;
  cards;
1 C male
2 C female
3 C male
5 C male
;
run;

data D;
  input patno source $ age;
  cards;
2 D 45
2 D 40
4 D 35
;
run;

data test3;
   merge C D;
run;


proc print data=test3;
run;
```

In Example 5.6, there are no messages printed in the SAS log signifying a problem. When reviewing the SAS log, you will see that all that is printed is the program itself, with none of the NOTEs that are usually created when a program runs. The reason for this is that the unbalanced quote in the TITLE statement causes the entire program to be part of the TITLE statement ❶. If the quoted string is less than 200 characters, SAS will not make any mention of the unbalanced quote. SAS will run the program and expect the other quote to come. But if the string is over 200 characters, SAS will give you a NOTE stating that the character string is over 200 characters and you may have an unbalanced quote.

5.2.6 *Variable Uninitialized and Variable Not Found*

When SAS is unable to find one or more of the variables specified in the DATA step, SAS will print the "Variable Uninitialized" or the "Variable Not Found" message. SAS will continue to run, creating the variable and setting its values to missing for each observation. SAS places all variables in the DATA step into the PDV. To prevent getting the message "Variable Uninitialized" or "Variable Not Found," make sure the variable exists in the data set, is not misspelled, and is initialized with a value. You may also want to make sure the correct data set is specified and to specify the SAS statement to create the variable before any SAS statement that uses the new variable.

Example 5.7   Missing variable in DATA step.

tests.txt

```
   Tim math 9 10
   Sally math 10 10
   John math 8 6
-----------------------------------------------------------
data tests;
   infile 'tests.txt';
   input name $ class $ quiz1 quiz2;
   total=quiz1+quiz2+quiz3; ❶
run;
```

Log

```
1     data tests;
2         infile 'tests.txt';
3         input name $ class $ quiz1 quiz2;
4         total=quiz1+quiz2+quiz3;
5     run;

NOTE: Variable quiz3 is uninitialized. ❷
NOTE: The infile 'tests.txt' is:
      File Name=C:\Documents and Settings\Owner\tests.txt,
      RECFM=V,LRECL=256

NOTE: 3 records were read from the infile 'tests.txt'.
      The minimum record length was 15.
      The maximum record length was 19.
```

```
NOTE: Missing values were generated as a result of
      performing an operation on missing values ❸.
      Each place is given by: (Number of times) at
      (Line):(Column).
      3 at 4:20
NOTE: The data set WORK.TESTS has 3 observations and 6
      variables.
NOTE: DATA statement used:
      real time            0.40 seconds
      cpu time             0.09 seconds
```

In Example 5.7, SAS is unable to find the *quiz3* variable ❶; therefore, SAS prints a message that the variable is uninitialized ❷. SAS will then create the variable and set all of its values to missing, causing the next message about the missing values ❸. Because the SUM function was not specified to SUM variables *quiz1*, *quiz2*, and *quiz3*, the total variable will contain only missing values in the *quiz3* variable. See Example 3.7.

A missing variable is more serious when it occurs within a SAS procedure. There are many reasons why missing variables occur, but some of the more common reasons are that the variable name is misspelled, the variable has been dropped, the wrong data set is being used, or the variable is being used before it is created. If a variable does not exist in an SAS procedure, SAS will create an ERROR message and stop executing.

Example 5.8 Missing variable in a procedure.

```
data tests;
   infile 'tests.txt';
   input name $ class $ quiz1 quiz2;
  run;

proc print data=tests;
   var name class quiz1 quiz2 quiz3;
run;
```

Log

```
10
11    proc print data=tests;
12       var name class quiz1 quiz2 quiz3;
ERROR: Variable QUIZ3 not found. ❶
13    run;
```

```
NOTE: The SAS System stopped processing this step because
      of errors.
```
NOTE: PROCEDURE PRINT used:
```
      real time             0.20 seconds
      cpu time              0.01 seconds
```

In Example 5.8, the ERROR message is created and SAS stops executing since the variable *quiz3* is missing ❶ in the *tests* data set. This type of error is generally due to a variable accidentally being dropped in the DROP statement or the variable not being included in the KEEP statement.

5.2.7 MERGE Statement Has More Than One Data Set with Repeats of BY Variables

"MERGE statement has more than one data set with repeats of BY variables" is a NOTE that can appear in the SAS log. This message is created when SAS encounters a problem when executing a MERGE statement with the BY statement. **When SAS merges data sets where two or more of the data sets in the MERGE statement contain multiple observations that have the same BY values, SAS does not know which observations to merge from one data set to another.** Basically, there is not a unique match between data sets in that multiple records from one data set are matched with multiple records from another data set. You can correct this problem by including variables in the BY statement that uniquely identify the records in each data set. As discussed in Chapter 1, Section 1.6, SAS performs a match merge and creates a data set that contains all of the unique combinations of the BY variable. Although SAS will merge the data sets, the results will be unpredictable and may be incorrect.

Example 5.9   MERGE statement has more than one data set with repeats of BY variables.
```
data tests;
   input name $ class $ quiz_score ;
   cards;
   Tim math 9
   Tim history 7
   Tim science 7
   Sally math 10
   Sally science 7
   Sally history 5
   John math 8
```

```
   John  history 7
   John    science  6 ❷
;
run;

data exam;
   input name $ class $ exam_score ;
   cards;
   Tim history 93
   Tim science 86
   Tim math 85
   Sally history 93
   Sally math 78
   Sally science 84
   John math 94
   John science 84 ❸
   John history 73

;
run;

proc sort data=tests;
   by name;
run;

proc sort data=exam;
   by name;
run;

data grade;
   merge tests
         exam;
   by name;
run;

proc print data=grade;
run;
```

Log (Partial)

```
128
129   data grade;
130      merge tests
```

```
131                exam;
132       by name;
133    run;


NOTE:  MERGE statement has more than one data set with
       repeats of BY values. ❶
NOTE:  There were 9 observations read from the data set
       WORK.TESTS.
NOTE:  There were 9 observations read from the data set
       WORK.EXAM.
NOTE:  The data set WORK.GRADE has 9 observations and 4
       variables.
NOTE:  DATA statement used:
       real time              0.09 seconds
       cpu time               0.06 seconds
```

Output

Obs	name	class	quiz_ score	exam_ score
1	John	math	8	94
2	John	science	7	84 ❹
3	John	history	6	73
4	Sally	history	10	93
5	Sally	math	7	78
6	Sally	science	5	84
7	Tim	history	9	93
8	Tim	science	7	86
9	Tim	math	7	85

In Example 5.9, the output data set looks good. The nine observations in each of the data sets combine together in the output data set. But, if you review the SAS log, you will notice the NOTE "MERGE statement has more then one data set with repeats of BY values" ❶. Upon closer examination of the output data set you will notice that there are data problems. In the *tests* data set, the observation for John has his science quiz score as 6 ❷ and the exam score in the exam data set is 84 ❸. In the combined data set, the science quiz score is 7 with the exam score of 84. ❹ The difference is because there are repeats of BY variables; therefore SAS will try to merge the data by taking the first observation of each BY variable and match them together and then the second and so on. The result is that the last observation in the *tests* data set for John contains his science quiz score and the last observation in the *exam* data set contains his history score. When you combine the observations, the *class* variable becomes science since SAS will use the second data set value when the variables are the same between the two data sets. The quiz and exam values correspond to the values in the input data sets. The way to correct this problem is to make sure that you have unique observations in your BY statement. In this example, the data sets should be merged by *name* and *class*, so each resulting observation is unique. Remember to sort by the same BY variables.

5.2.7.1 Corrected Data Set

```
                The SAS System

                            quiz_    exam_
Obs    name      class      score    score

 1     John      history      7        73
 2     John      math         8        94 ❺
 3     John      science      6        84
 4     Sally     history      5        93
 5     Sally     math        10        78
 6     Sally     science      7        84
 7     Tim       history      7        93
 8     Tim       math         9        85
 9     Tim       science      7        86
```

In the corrected data set, the values for John's math quiz and exam score are correct because they are consistent with the *tests* and *exam* data sets ❺.

5.2.8 Variable Type Conflict

You may encounter a variable type conflict if the BY variables have different variable types when merging data sets. Note that SAS will issue a NOTE and WARNING when there is a mismatch of data types when specifying the MERGE, SET, or UPDATE statements. **Make sure common variables have the same attributes.** If, however, the common variables have the same name and variable type but different lengths, then SAS does not issue a NOTE or WARNING even though SAS uses the first data set's variable length, even if it is the shorter length. You may need to convert the variable from one of the data sets to match the other data set. Example 3.9 in Chapter 3 shows how to apply the PUT() or INPUT() functions to convert the variable type.

Example 5.10 *Patno* variable as numeric and character.

```
data C;        ❶
  input patno $ source $ gender $;
  cards;
1 C male
2 C female
3 C male
5 C male
;
run;

data D;   ❷
  input patno source $ age;
  cards;
2 D 45
2 D 40
4 D 35
;
run;

proc sort data=C;
  by patno;
run;

proc sort data=D;
  by patno;
run;

data test4;
     merge C D;
     by patno;
run;

proc print data=test4;
run;
```

Output

```
62     data test4;
63     merge C D;
ERROR: Variable patno has been defined as both character
       and numeric. ❸
```

```
64    by patno;
65    run;
```

```
NOTE: The SAS System stopped processing this step because
      of errors.
WARNING: The data set WORK.TEST4 may be incomplete.  When
         this step was stopped there were 0
         observations and 4 variables.
WARNING: data set WORK.TEST4 was not replaced because this
         step was stopped.
NOTE: DATA statement used:
```

In Example 5.10, the *patno* variable has been defined as both a character variable (data set C) ❶ and a numeric variable (data set D). ❷ This will cause SAS to print an ERROR in the SAS log and stop the processing when the two data sets are being merged ❸. In order to correct the ERROR, the *patno* variable should be changed so that it has the same variable type between the two data sets that are being merged.

5.2.9 BY Group Processing

SAS generates an ERROR when specifying an SAS procedure with the BY statement without first presorting the data set.

Example 5.11   Using the MEANS procedure without the SORT procedure.

```
data tests;
   input name $ class $ test_score ;
   cards;
   Tim math 9
   Tim history 7
   Tim science 7
   Sally math 10
   Sally science 7
   Sally history 5
   John math 8
   John history 7
   John   science 6
 ;
 run;
```

```
proc means data=tests n mean noprint;
   var test_score;
   by name; ❶
      output out=test_stat n(test_score)= test_score_n
   mean(test_score)=test_score_mean;
run;


proc print data=test_stat;
run;
```

Log

```
16    proc means data=tests n mean noprint;
17        var test_score;
18        by name;
19        output out=test_stat n(test_score)= test_score_n
20        mean(test_score)=test_score_mean;
21    run;
```

ERROR: Data set WORK.TESTS is not sorted in ascending
       sequence. The current by-group has name =
Tim and the next by-group has name = Sally.
NOTE: The SAS System stopped processing this step because
      of errors.
NOTE: There were 4 observations read from the data set
      WORK.TESTS.
WARNING: The data set WORK.TEST_STAT may be incomplete.
         When this step was stopped there were
            0 observations and 5 variables.
NOTE: PROCEDURE MEANS used:

Example 5.11 shows the ERROR message produced when SAS expects the data set to be sorted by the *name* variable before processing the MEANS procedure ❶. Make sure to presort the data if required by the SAS procedure.

5.3   Examining and Resolving Data Errors

Data errors can also be difficult to find since they generally appear in the SAS log as NOTEs and are easily missed until you print your data. Table 5.2 shows common types of data errors and how to resolve them. **It might be helpful to use the PUT statement with conditional execution statements to display unexpected data or the record number of data issues.**

Table 5.2   Common Data Errors and How to Resolve Them

Data Error	Resolution
Missing values	Apply conditional logic or use function such as SUM() or MEAN() to exclude missing values
Numeric to character and character to numeric conversions	Use PUT() and INPUT() functions to convert variable types
Invalid data	Confirm data assumptions or adjust INPUT statement to read data correctly
Character field truncated	Specify an ATTRIB or LENGTH statement to capture all data

5.3.1   Missing Values

The "Missing values were generated" NOTE is created in the SAS log when SAS is unable to compute the value of a new variable since one of the variables used in the creation of the new variable is missing. Although this may not be a problem, these NOTEs in the SAS log should be cleaned up. One of the methods that can be used to avoid these NOTEs in the SAS log is to have the new variable created only when the other variables are not missing. This can be done in two ways. If there is only one variable that is used to create the new variable, then an IF condition can be specified so that the calculation creating the new variable is only executed when the calculation variable is not missing. If there is more than one variable in the calculation then the NMISS() function can be specified to condition the creation of the new variable. The NMISS() function returns the number of missing values in the variables specified. In addition, use functions such as SUM() and MEAN() to calculate the sum and mean of variables instead of manually constructing the SAS statement. When using the SUM() and MEAN() functions, SAS automatically ignores missing values. See Example 3.8 and Table 3.2 for more information on SAS functions.

Example 5.12   Missing values in calculations.

```
Data bday;
   input name $ birthdt date9. ;
   format birthdt date9.;
   cards;
   Tim 12MAR1970
   Adam 25APR1968
```

```
   Brian  . ❸
   Sally 01JUN1965
   Michelle 22JAN1973


 ;
 run;

 data age;
    set bday;
    format todaydt date9.;
    todaydt=today();
    age=(todaydt-birthdt+1)/365.25;
    if nmiss(todaydt,birthdt)=0 then age2=(todaydt-
 birthdt+1)/365.25;  ❹
     if age = . then put '** Problem Data **: ' age =
 todaydt =   birthdt = ;  ❻
 run;

 proc print data=age;
 run;
```

Log (Partial)
```
 130
 131   data age;
 132      set bday;
 133      format todaydt date9.;
 134      todaydt=today();
 135      age=(todaydt-birthdt+1)/365.25;
 136      if nmiss(todaydt,birthdt)=0 then age2=(todaydt-
          birthdt+1)/365.25;
 137   run;


 ** Problem Data **: age=. todaydt=24AUG2004 birthdt=.  ❼
 NOTE: Missing values were generated as a result of
       performing an operation on missing values.
       Each place is given by: (Number of times) at
       (Line):(Column).  ❶
       1 at 135:15  ❷
 NOTE: There were 5 observations read from the data set
       WORK.BDAY.
 NOTE: The data set WORK.AGE has 5 observations and 5
       variables.
```

```
NOTE:  DATA statement used:
       real time              0.07 seconds
       cpu time               0.04 seconds
```

Output

Obs	name	birthdt	todaydt	age ❺	age2
1	Tim	12MAR1970	21JUN2004	34.2806	34.2806
2	Adam	25APR1968	21JUN2004	36.1588	36.1588
3	**Brian**	**.**	**21JUN2004**	**.**	**.**
4	Sally	01JUN1965	21JUN2004	39.0582	39.0582
5	Michelle	22JAN1973	21JUN2004	31.4141	31.4141

In the SAS log in Example 5.12, the missing value NOTE is displayed due to missing values in the creation of the *age* variable ❶. The NOTE also shows you the place and the number of times a missing value was created. As you can see, the missing value was generated once at line 135 ❷. If you look back at the data you will notice that one observation did not have a value for *birthdt* ❸. Since the missing value NOTE is due to the data and not an indication of a problem, it would be better to remove the message. The creation of the *age2* variable is one way to create an *age* variable without getting a missing value NOTE. ❹ The NMISS() function is used to check the calculation variables to make sure that none of these variable's values are missing. If none of the calculation variables have missing values, then the *age2* variable will be calculated. If you look at the output, you will notice that the results for *age* and *age2* are the same ❺, but the creation of the *age2* variable will prevent generating the missing value NOTE. Note that by specifying the PUT statement for the condition age = . will display the *todaydt* and *birthdt* values for the same record in the SAS log ❻ ❼.

5.3.2   Numeric to Character and Character to Numeric Conversions

Conversion NOTEs in the SAS log are generally due to character or numeric variables accidentally being mixed up in the same SAS statement. If you have character variables specified in calculations, SAS will convert the character variable to numeric and continue running the program. Other examples include setting a numeric variable equal to a character variable or using a numeric variable in a character function. SAS will print a NOTE in the SAS log showing the line and the column where the conversion occurred. In order to avoid these conversion messages in the SAS log, you and not SAS should do the variable conversion. There are two functions that you can use to convert variable types.

The INPUT() function is used to convert character to numeric variables and the PUT() function is used to convert numeric to character variables. As discussed in Chapter 3, Section 3.6, in either case, the format or informat must be numeric.

Example 5.13   Character to numeric conversion.

```
data tests;
   input name $ class $ quiz1 $   quiz2 $; ❷
   cards;
   Tim math 9 10
   Sally math 10 10
   John math 8 6
;
run;

data total;
   set tests;
   total=quiz1+quiz2;
   total2=input(quiz1,2.) + input(quiz2,2.); ❸
run;

proc print data=total;
run;
```

Log (Partial)

```
177
178   data total;
179      set tests;
180      total=quiz1+quiz2;
181      total2=input(quiz1,2.) + input(quiz2,2.);
182   run;

NOTE: Character values have been converted to numeric
      values at the places given by: (Line):(Column).
      180:9     180:15 ❶
NOTE: There were 3 observations read from the data set
      WORK.TESTS.
NOTE: The data set WORK.TOTAL has 3 observations and 6
      variables.
```

```
NOTE:  DATA statement used:
       real time           0.06 seconds
       cpu time            0.04 seconds
```

Output

Obs	name	class	quiz1	quiz2	total ❹	total2
1	Tim	math	9	10	19	19
2	Sally	math	10	10	20	20
3	John	math	8	6	14	14

In the SAS log in Example 5.13, you can see that at line 180, columns 9 and 15, SAS converted a character value to numeric ❶. From reviewing the INPUT statement, you can see that the *quiz1* and *quiz2* variables were created as character variables ❷. This is a data issue and not a problem so it is better to remove the NOTE. The *total2* calculation uses the INPUT() function to convert the variables to numeric before they are added together ❸. Using these functions allow you to convert the variables yourself and remove the conversion NOTE. The numeric format of 2. is applied to each variable. Note that although in this example, SAS correctly converted the numeric data stored in the character variable to generate the same results as the variable *total2*, it is always better to convert the variable yourself ❹.

Sometimes a data set might have invalid data values. For example, the character dates entered in a data set may contain partial dates such as Jan2004 or 2003. If an attempt is made to convert these dates into numeric values, the SAS log will contain error messages. The use of optional question mark (?) and double question mark (??) format modifiers suppress the printing of both the error messages and the input lines when invalid data values are read. The "?" modifier suppresses the invalid data message. The "??" modifier also suppresses the invalid data message and, in addition, prevents the automatic variable _ERROR_ from being set to 1 when invalid data is read.

Example 5.14   Inconsistent raw date formats.

```
data demo;
   input name $ birthdtv $9.;
   datalines;
Jack  01Jan2001
John  Dec2004 ❶
;
run;
```

```
data newdemo;
   set demo;
   birthdt=input(birthdtv,date9.);
run;
proc print data=newdemo;
run;
```

Log (Partial)

```
NOTE: Invalid argument to function INPUT at line 11 column
11. ❷
name=John birthdtv=Dec2004 birthdt=. _ERROR_=1 _N_=2
NOTE: Mathematical operations could not be performed at
      the following places. The results of the
      operations have been set to missing values.
      Each place is given by: (Number of times) at
      (Line):(Column).
      1 at 11:11
NOTE: There were 2 observations read from the data set
      WORK.DEMO.
NOTE: The data set WORK.NEWDEMO has 2 observations and 3
      variables.
NOTE: DATA statement used:
      real time              0.32 seconds
      cpu time               0.13 seconds
```

Output

```
Obs   name      birthdtv      birthdt

1     Jack      01Jan2001     14976
2     John      Dec2004          .
```

In Example 5.14, the second observation for *birthdtv* is only a partial date ❶. So when SAS tries to read this value into the *birthdt* variable, an "Invalid argument to function INPUT()" NOTE is printed in the SAS log ❷. In Example 5.15, the "??" modifier is used with the INPUT() function to prevent the NOTE.

Example 5.15   "?" and "??" modifiers to prevent invalid data.

```
data demo;
   input name $ birthdtv $9.;
   datalines;
```

```
Jack 01Jan2001
John Dec2004
;
run;

data newdemo;
   set demo;                        ❶
   birthdt=input(birthdtv,?? date9.);
run;

proc print data=newdemo;
run;
```

Log (Partial)

NOTE: **There were 2 observations read from the data set**
      WORK.DEMO.
NOTE: **The data set WORK.NEWDEMO has 2 observations and 3**
      variables.
NOTE: DATA statement used:
      real time   0.04 seconds
      cpu time   0.04 seconds

Output

Obs	name	birthdtv	birthdt
1	Jack	01Jan2001	14976
2	John	Dec2004	. ❷

In Example 5.15, because the "??" modifier is used, ❶ there are no messages printed in the SAS log due to invalid data. The observation that had the partial date will have a missing value for the *birthdt* variable. ❷ Care must be used when the "?" and "??" modifiers are specified since no messages will be printed in the SAS log when there is invalid data.

5.3.3 *Invalid Data*

SAS generates the "Invalid Data" NOTE when it encounters unexpected data while reading data using the INPUT statement. When SAS encounters invalid data, it will set the problematic variable to missing for that observation and then print the "Invalid Data" NOTE. There are many reasons why the "Invalid Data" NOTE is created, but some of the more common reasons are listed below:

- Forgetting to specify a variable is character (SAS assumes that it is numeric, by default)
- Incorrectly specifying columns to produce embedded spaces in numeric data or character values for a numeric variable
- Using the LIST INPUT statement to read data with two periods in a row without a space in between
- Using the LIST INPUT statement to read missing values without them being marked as periods, causing SAS to read the data for the next variable
- Using the letter O instead of the number zero
- Trying to read special characters such as a carriage return
- Reading invalid dates with a date informat (such as "January 0")
- Using the wrong informat

Sometimes you may need to create a test data set containing a variety of data values to confirm conditional executions within the data set. If possible, you should take advantage of SAS's new feature for preventing data errors with integrity constraints in Version 8.2. See Chapter 6, Section 6.2 for more information.

Example 5.16 Invalid data.

```
 data bday;
    input name $ 1-9 birthdt date9.;
    format birthdt date9.;
    cards;
    Tim      12MAR1970
    Adam     25APR1968
    Brian    .
    Sally    01JUN1965
 ❸ Michelle 22JAN1973
 ;
 run;
 proc print;
 run;
```

Log

```
 210   data bday;
 211      input name $ 1-9 birthdt date9.;
```

```
212      format birthdt date9.;
213      cards;

NOTE: Invalid data for birthdt in line 218 10-18. ❶
RULE:     ---+---1---+---2---+---3---+---4---+---5---+---
218           Michelle 22JAN1973 ❷
name=Michell birthdt=. _ERROR_=1 _N_=5
NOTE: The data set WORK.BDAY has 6 observations and 2
      variables.
NOTE: DATA statement used:
      real time             0.07 seconds
      cpu time              0.04 seconds



220   ;
221   run;
```

Output

```
Obs   name      birthdt

1     Tim       12MAR1970
2     Adam      25APR1968
3     Brian         .
4     Sally     01JUN1965
5     Michell       . ❹
```

When SAS creates the "Invalid data" NOTE, SAS informs you which variable has the problem. In Example 5.16, the *birthdt* variable has the invalid data ❶. SAS will also inform you of the raw data line and column causing the problem: line 218, column 10–18. SAS will then print a ruler that indicates the number of the columns. On the ruler, the "1" indicates the tenth column. After the ruler, SAS will output the row of data that had the problem ❷. As you can see, the value for *name* actually started in column 3 instead of column 1 due to two spaces in the data ❸ and extended past the columns that were specified for the variable. For the last observation SAS tried to read the letter "e" of "michelle" into the *birthdt* variable. Since SAS expects a date formatted value for the *birthdt* variable starting from column ten, it assigns a missing value for that record ❹. Note that SAS will give the "Invalid data" message if the data is not in the format that SAS expects. Make sure all dates are in the same expected format. SAS will also display two automatic variables, _ERROR_ and _N_. If there is a data error, then the _ERROR_ variable will have a value of 1. The _N_ variable displays the number of records that SAS has read in the current DATA step. See Chapter 1, Section 1.2 for correctly reading in data using the INPUT statement.

5.3.4 Character Field Truncated

This type of error does not create any ERROR, WARNING, or NOTE in the SAS log. You will know that you have this problem only when you print your data and find that the end of a character variable has been truncated. As discussed in Chapter 1, Section 1.2, the length of a character variable is set when SAS first encounters the variable. Typically this will occur in an INPUT statement or assignment statement. If you use the LIST INPUT statement, the length of the character variable is 8 bytes by default. With the COLUMN INPUT statement, it is the number of columns that you specify. With the FORMATTED INPUT statement, the length of the character variables is the length of the informat. Note that the length of numeric variables in all three types of INPUT statements are still the default 8 bytes unless specified with a LENGTH or ATTRIB statement. If a new variable is created in an assignment statement, SAS will assign the length of the variable based on the first occurrence of the variable's value. See Section 1.2 for more information on INPUT statements.

Example 5.17   Character field truncated.
```
data tests;
   input name $ class $ quiz1 quiz2 ;
   cards;
   Tim math 9 10
   Sally math 10 10
   John math 8 6
;
run;

data level;
   set tests;
   total=quiz1 + quiz2;
   if total<16 then level='Low'; ❶
   else level='High';
run;
```

```
proc contents data=level;
run;
proc print data=level;
run;
```

Output

```
     -----Alphabetic List of Variables and Attributes-----

          #      Variable      Type      Len          Pos

          2       class        Char       8            32
          6       level        Char       3      ❷     40
          1       name         Char       8            24
          3       quiz1        Num        8             0
          4       quiz2        Num        8             8
          5       total        Num        8            16
```

Obs	name	class	quiz1	quiz2	total	level
1	Tim	math	9	10	19	Hig ❸
2	Sally	math	10	10	20	Hig
3	John	math	8	6	14	Low

In Example 5.17, since the new variable *level* was not previously specified before the assignment statement, SAS determines the length of 3 from the value Low and the variable type of character because of the quotes. Since SAS defines the variable with a length of 3, ❷ it will only store the first three letters of the High value. The last letter "h" is truncated.

The PRINT procedure shows the truncated values for the variable level ❸. The best way to solve this problem is to make sure that you use an ATTRIB or LENGTH statement with any new variables that you create instead of letting SAS determine the length of the variable for you. See Example 3.3 for more information on ATTRIB statement.

5.4 Program Debugging Techniques

Although there is not a standard process for debugging programs, this section helps guide you through the debugging process. **After you have run your program, it is a good practice to look at your SAS log and search for the text ERROR or WARNING.** If you find an ERROR, you need to fix the problem to prevent the program from aborting. WARNINGs are less serious, in that your program will run but SAS is doing something you may not have wanted. Although your program will run with WARNINGs, these should also be corrected. In addition to searching for the text ERROR or WARNING you should also check for messages that could appear as NOTEs. Additional keywords that should also be searched for are: "uninitialized," "repeats of BY," "converted," and "missing values."

If you discover an ERROR or WARNING or one of the keywords above in your SAS log, use the following steps to identify the problem and fix your program.

1. Work backwards from the first ERROR or WARNING message. Some ERROR or WARNING messages will cause more than one message to appear in your SAS log. Finding the first message and correcting it may fix the other ERROR or WARNING messages.

2. When you have found the first ERROR or WARNING message but you don't know exactly what the problem is, check the syntax of your program. Common types of syntax problems are listed below:

   - Missing semicolon
   - Errors or omissions in the DATA step coding
   - Array subscript out of range
   - Missing END with a DO
   - Uninitialized variables
   - Misspelled keyword
   - DATA step statement in an SAS procedure or vice versa
   - Unmatched quote
   - Unmatched comment
   - Misspelled variable name
   - Wrong data set
   - Incorrect option for statement
   - Missing RUN statement

3. After you have corrected all of the ERROR and WARNING messages you should fix any of the NOTEs that contain the keywords that were searched for above.

Now that you have checked your SAS log, you need to confirm the results of your output. Always review your data; one of the biggest problems could be that you are not getting an ERROR message but you are getting the wrong results. Make sure that the numbers are consistent and reasonable and that they fulfill your requirements. Check the logic of your program to ensure that data is manipulated properly and that variables are created before they are used in calculations and functions. There are several SAS procedures listed below to help you review your data.

- PROC CONTENTS
  - Check the number of observations and variables available in the data set
  - Check variable attributes such as LENGTH and TYPE
- PROC PRINT
  - Before and after DATA steps, compare number of observations and variables
  - Use the OBS= option to print a subset of the data
  - Apply WHERE statement to check for specific conditions
  - Ensure that unintended duplicate records do not exist
- PROC FREQ
  - Best for categorical variables—see list of all unique values, including missing values
  - Create cross tabulation of variables to show all possible combinations of the variables' values
- PROC MEANS or PROC UNIVARIATE
  - Can be used to check for missing values
  - Check minimum and maximum values to determine out of range values
  - Check averages to determine if data are reasonable

In addition, resolve errors that are system-related. For example, if SAS cannot locate a file, confirm that the correct full path name is specified. Also, you may want to confirm resources and system options to prevent out of memory or space problems.

Along with SAS procedures, there are several methods for debugging the program within the DATA step, as shown in Table 5.3. The DATA step is useful to write messages to the SAS log at intermediate points within the DATA step.

The following list of SAS books and articles are excellent resources for writing, maintaining, and debugging SAS programs: *Professional SAS Programming Secrets, Mastering the SAS System, SAS Applications Programming: A Gentle Introduction, The Elements of SAS Programming Style, SAS Programming Conventions, The Little SAS Book: A Primer, The Next Step Integrating the Software Life Cycle with SAS Programming, Taming the Chaos: A Primer on the Software Life Cycle and Programming Standards, Building Quality into the SAS System, The Art of Testing Programs with an Emphasis on Larger Files, Validation for the Rest of Us, Debugging 101* and *Debugging SAS Programs: A Handbook of Tools and Techniques.*

Table 5.3 DATA Step Debugging Techniques

Statement	Description
`IF AGE < 0 OR AGE > 50 THEN PUT 'age is not correct: ' AGE 3.;`	Displays the value of out of range data.
`IF AGE < 0 OR AGE > 50 THEN PUT _INFILE_;`	Displays the last record accessed by an INPUT statement and all variables currently in the program data vector.
`IF AGE < 0 OR AGE > 50 THEN PUT _ALL_;`	Displays values of all variables.
`IF AGE < 0 OR AGE > 50 THEN PUT AGE = ;`	Displays AGE= value.
`IF AGE < 0 OR AGE > 50 THEN LIST;`	Displays the last record accessed by an INPUT statement with a column ruler. This helps identify incorrect INPUT statement column specifications.
`IF _ERROR_ THEN LIST;`	Displays all variables if error exists.
`IF AGEGRP='OLD' AND AGE < 19 THEN ERROR 'AGEGRP and AGE do not match';`	ERROR statement can be used not only to write messages to the log, but to cause the _ERROR_ flag to be set to 1, thereby forcing a dump of the input buffer.
`IF AGEGRP='OLD' AND AGE < 19 THEN _ERROR_ = 1; ELSE _ERROR_ = 0;`	Set the _ERROR_ flag to 1 to dump listing of all variables, else set to 0 to prevent display.
`MERGE C (IN=INC) D (IN=IND); BY PATNO; IF AGE < 19 THEN PUT 'Age < 19' INC= IND=;`	Use the IN= data set option to identify which data set(s) a particular observation came from during a MERGE, SET, or UPDATE operation.
`OPTIONS DETAILS ERRORS NOTES SYNTAXCHECK;`	System options to help debug programs. SYNTAXCHECK is a 9.1 system option.
`Enhanced Editor`	Enhanced editor can be used to change the color of keywords in SAS programs.
`DATA A / DEBUG; ... RUN;`	Invokes the debugger feature in SAS Display Manager.

Chapter 5. Handling Errors—Chapter Summary

Syntax Error (Compile Time Error)	*Resolution*
Missing semicolon	Add semicolon
Misspelled/missing keywords or incomplete SAS statement	Correct the keyword spelling or add expected SAS statement
WHERE clause operator requires compatible variables	Ensure both operator and variable types match
At least one W.D. format was too small to be displayed	Change format or increase variable length
Unbalanced quotes	Add missing quote

Non-Syntax Error (Execution Time Error)	*Resolution*
Variable uninitialized and variable not found	Make sure variable exists and is initialized if it is a new variable
MERGE statement has more than one data set with repeats of BY variables	Ensure BY variables uniquely separate records in each data set
Variable type conflict	Ensure common BY variables when merging data sets
BY group processing	Presort data sets before specifying the BY statement in DATA step and SAS procedures

Data Error	*Resolution*
Missing values	Apply conditional logic or use function such as SUM() or MEAN() to exclude missing values
Numeric to character and character to numeric conversions	Use PUT() and INPUT() functions to convert variable types
Invalid data	Confirm data assumptions or adjust INPUT statement to read data correctly
Character field truncated	Specify an ATTRIB or LENGTH statement to capture all data

Chapter 5.   Handling Errors—Chapter Questions

Question 1: Name at least two ways ERRORS or WARNING messages can be caused.

Question 2: Name at least two methods for debugging SAS programs.

Question 3: Is it possible to get the note "Invalid data" when using the SET statement in a DATA step?

Question 4: Do all syntax errors prevent a data set from being created?

Question 5: In general, when you see the NOTE "At least one W.D. format was too small for the number to be printed," does SAS incorrectly store the data or incorrectly display the data?

Question 6: When you see the NOTE "Missing values are generated," does SAS set all values of the new variable to missing?

Question 7: When you see the NOTE "Character values have been converted to numeric values," does SAS convert the character values to zeros?

Question 8: What does SAS do when it encounters a missing variable?

Question 9: What is the best way to resolve the following SAS NOTE: "Merge statement has more than one data set with repeats of BY variable"?

Question 10: What are the two possible cases for getting the following ERROR message: "WHERE clause operator requires compatible variables"?

Chapter 6
Version 8.2 and Version 9.1 Enhancements

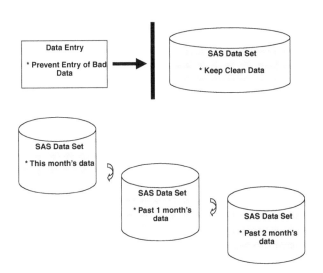

PROCESS FLOW DIAGRAM		
PHASE	**CHAPTER. DESCRIPTION**	
INPUT:	1. Accessing Data	
	2. Creating Data Structures	
PROCESS:	3. Managing and Summarizing Data	
OUTPUT:	4. Generating Reports	
HANDLING ERROR:	5. Diagnosing and Correcting Errors	
V8.2/9.1:	6. Integrity Constraints, Generation DS, Audit Trials	

6.1 Introduction

Chapter 6 introduces some of the new features in Version 8.2 and Version 9.1. Some of the new features in Version 8.2 are the ability to concatenate libraries and catalogs, new benefits in defining variable attributes, data integrity constraints, audit trails, and generation data sets. The Version 9.1 features that will be discussed are multiprocessing capabilities, new ODS formatting, new functionality with SAS procedures, and new SAS functions. It is not within the scope of this book to show all Version 8.2 and Version 9.1 enhancements. A sample of some of the important enhancements are reviewed.

6.2 Version 8.2 Enhancements

6.2.1 Concatenating Libraries and Catalogs

Syntax:
LIBNAME *libref* *<engine>* (*library-specification*-1 *<. . . library-specification-n>*) *< options >*;
CATNAME'*<libref.>* *catref* (*libref-1.catalog-1 ...libref-n.catalog-n*);

You have seen the use of the LIBNAME statement in earlier chapters to associate a single library to a libref. The syntax for specifying multiple libraries is very similar. The syntax starts the same, with the LIBNAME keyword followed by the libref. The libraries are then listed in parentheses. The full path of the library can be listed or a libref can be used. The reason for using multiple libraries in a single libref is that it provides greater flexibility for prioritizing and organizing SAS files. SAS will search through multiple libraries using a single libref name without you having to specify separate LIBNAME statements. When SAS reads a file from the library, it locates the first occurrence of the file when searching the concatenation library. When SAS writes a file to a library, it writes to the first library in the concatenation library.

When there are both Version 6.12 and Version 8.2 files in the same directory, specify the V6 engine type on the LIBNAME statement to access Version 6.12 files or else they will not be read. If you have files with mixed versions within the same directory, it is best to create separate LIBNAME statements for each version.

Catalogs are files that contain different types of application information such as window definitions, help windows, formats, informats, macros, or graphics output. The search order for catalogs is similar to that for the library concatenation catalog. Catalog entries are written to the first catalog listed in the concatenation. When reading catalog entries, SAS searches the catalog in the first catalog listed before going to the other catalogs.

Example 6.1   Concatenating libraries.

```
libname mylib ('c:\myproject1' 'c:\myproject2');
libname alllib (mylib, 'c:\otherprojects');
```

Log

```
7    libname mylib ('c:\myproject1' 'c:\myproject2');
NOTE: Libref MYLIB was successfully assigned as follows:
      Levels:                    2
      Engine(1):                 V8
      Physical Name(1):  c:\myproject1
      Engine(2):                 V8
      Physical Name(2):  c:\myproject2
8    libname alllib (mylib, 'c:\otherprojects');
NOTE: Libref ALLLIB was successfully assigned as follows:
      Levels:                    3
      Engine(1):                 V8
      Physical Name(1):  c:\myproject1
      Engine(2):                 V8
      Physical Name(2):  c:\myproject2
      Engine(3):                 V8
      Physical Name(3):  c:\otherprojects
```

In Example 6.1, the *myproject1* and *myproject2* libraries are assigned to the *mylib* libref. In the second LIBNAME statement, the *mylib* libref is concatenated with the *otherprojects* library to create the *alllib* libref. When applying the *alllib* libref, SAS searches for the data set in the priority as specified in the LIBNAME statement. SAS will first look in the *myproject1* library, followed by the *myproject2* library and then the *otherprojects* library. In other words, if a data set exists in multiple libraries, the first occurrence will be the data set that is used. In addition, SAS will write files to the first library, which is *myprojects1*.

6.2.2   *Defining Variable Attributes*

A key benefit of Version 8.2 is to take advantage of specifying up to 32 characters in library and file references, formats, informats, variables, data sets, macro variables, and macro names. Previous versions of SAS allowed only up to eight characters. When creating variables, both upper and lower case characters can be used, but when they are stored in SAS they are stored as upper case. This means that you cannot create two variables with the same name but different cases.

Another benefit of Version 8.2 is the ability to have longer character values and labels. Character data can be up to 32,000 characters in length. The previous limit in earlier versions of SAS was 200 characters. Labels for both data sets and variables have also been increased; now labels can be up to 256 characters in length.

The VALIDVARNAME= system option can be helpful in migrating from Version 6.12 to Version 8.2. If VALIDVARNAME=V6, then the variables are uppercased and can only be eight characters in length. If the variables follow these rules, then the program will be able to be run in both versions. If the variables do not follow these rules then the program will not run in Version 6.12. If VALIDVARNAME=V6 and the program is executed in Version 8.2, then an ERROR message will be printed in the SAS log stating that the variable is an invalid SAS name. If VALIDVARNAME=V7, then the variable names can be up to 32 characters in length and can be displayed as mixed case. VALIDVARNAME=UPCASE is the same as VALIDVAR-NAME=V7 except all variables are stored as uppercase, as in earlier versions of SAS.

Example 6.2 Defining character attributes (VALIDVARNAME=V7).

```
options validvarname=v7;

data tests;      ❶
   input Name $ CLASS $ quiz_score ;
   cards;
   Tim math 9
   Tim history 7
   Tim science 7
   Sally math 10
   Sally science 7
   Sally history 5
   John math 8
   John history 7
   John   science 6
;
run;

data tests;
   set tests;
   where name='John';
run;      ❷
```

```
proc contents data=tests;
run;

proc print data=tests;
run;
```

Output

```
                The CONTENTS Procedure

     ----Alphabetic List of Variables and Attributes-----

            #    Variable      Type    Len   Pos

            2    CLASS         Char     8     16
            1    Name ❸        Char     8      8
            3    quiz_score    Num      8      0


                              ❹         quiz_
            Obs   Name     CLASS       score

             1    John     math           8
             2    John     history        7
             3    John     science        6
```

In Example 6.2, the *Name* variable is created as a mixed case variable name ❶, but in the second DATA step the WHERE clause specifies the variable as all lower case ❷. When displaying the data, SAS will output the variable as it was created ❸. **SAS, however, will store the variable name internally as upper case, so that variables that are the same name but different cases will be the same variable.** Note that because mixed case variable names are possible in Version 8.2, you can see the effect this has when displaying the data. Each variable—*Name, CLASS,* and *quiz_score*—is displayed exactly as it was specified in the INPUT statement ❹.

Example 6.3   Defining character attributes (VALIDVARNAME=V6).

```
options validvarname=v6;  ❶

data tests;
   input Name $ CLASS $ quiz_score ;
```

```
   cards;
   Tim math 9
   Tim history 7
   Tim science 7
   Sally math 10
   Sally science 7
   Sally history 5
   John math 8
   John history 7
   John science 6
;
run;

data tests;
   set tests;
   where name='John';
run;

proc contents data=tests;
run;

proc print data=tests;
run;
```

Log

```
117   options validvarname=v6;
118
119   Data tests;
120     input Name $ CLASS $ quiz_score ;
ERROR: The variable named QUIZ_SCORE contains more than
       8 characters. ❷
121     cards;

NOTE: The SAS System stopped processing this step because
      of errors.
WARNING: The data set WORK.TESTS may be incomplete. When
         this step was stopped there were 0
         observations and 2 variables.
WARNING: data set WORK.TESTS was not replaced because this
         step was stopped.
```

```
NOTE: DATA statement used:
      real time                     0.04 seconds
      cpu time                      0.03 seconds
```

Since in Example 6.3 the option VALIDVARNAME= is set to V6, ❶ SAS gives an ERROR ❷ message stating that the variable *quiz_score* is more than eight characters. The VALIDVARNAME=V6 value is useful when you are migrating your version of SAS and you need to be able to run your programs in both versions.

6.2.3 Preventing Data Errors with Integrity Constraints

Syntax:
PROC DATASETS;
  MODIFY <data set name>;
  INTEGRITY CONSTRAINT CREATE
  <constraint name> = <constraint>   /* **not null, unique, check** */
  MESSAGE = 'message string';
QUIT;

Integrity constraints are helpful for enforcing the consistency and correctness of your data. The constraints are created by using the DATASETS procedure. In the DATASETS procedure the MODIFY statement must be specified along with the data set name that you want to prevent data entry errors. **The INTEGRITY CONSTRAINT CREATE statement instructs SAS to create a constraint.** After the INTEGRITY CONSTRAINT CREATE statement the name of the constraint is specified along with the constraint. Some of the most used constraints are the NOT NULL, UNIQUE, and CHECK constraints. The NOT NULL constraint will check the specified variable to make sure no values for the variable are missing; the UNIQUE constraint will check to make sure that all values of a variable are unique. Finally, the CHECK constraint is used with the WHERE statement as a validity check with respect to data values and data ranges. **The constraints can be created for both numeric and character variables.** The DATASET procedure is then ended with the QUIT statement.

Example 6.4   Preventing data errors with integrity constraints.

```
options validvarname=v7;

data tests;
   input name $ class $ quiz_score ;
   cards;
```

```
   Tim math 9
   Steve history 7
   James science 7
   Sally math 10
   Michelle science 7
   Adam history 5
   Doug math 8
   Lynn history 7
   Dave science 6
;
run;

proc datasets;
   modify tests;
   integrity constraint create
      req_quiz_score = not null(quiz_score)
      message = 'quiz_score is REQUIRED';

   integrity constraint create
      ok_class = check(where=(class in ('math' 'history'
'science')))
      message = 'Class must be math, history or science';

   integrity constraint create
      unique_name=unique(name)
      message = 'name must be unique';
quit;

proc contents data=tests;
run;
```

Log (Partial)

```
Proc Datasets;

                                        -----Directory-----

            Libref:         WORK
            Engine:         V8
            Physical Name: C:\DOCUME~1\Owner\LOCALS~1\
Temp\SAS Temporary Files\_TD3104
```

```
            File Name:      C:\DOCUME~1\Owner\LOCALS~1\
Temp\SAS Temporary Files\_TD3104
```

#	Name	Memtype	File Size	Last Modified
1	TESTS	DATA	5120	05JUL2004:14:50:01

```
254    modify tests;
255     integrity constraint create
256        req_quiz_score = not null(quiz_score)
257        message = 'quiz_score is REQUIRED';
NOTE:  Integrity constraint req_quiz_score defined. ❶
258
259     integrity constraint create
260        ok_class = check(where=(class in ('math'
           'history' 'science')))
261        message = 'Class must be math, history or
           science';
NOTE:  Integrity constraint ok_class defined. ❷
262
263     integrity constraint create
264        unique_name=unique(name)
265        message = 'name must be unique';
NOTE:  Integrity constraint unique_name defined. ❸
266
267   quit;
```

Output

```
-----Alphabetic List of Integrity Constraints-----
```

#	Integrity Constraint	Type	Where Variables Clause	User Message
1	ok_class	Check	class in ('history', 'math', 'science')	Class must be math, history or science
2	req_quiz_ score	Not Null	quiz_score ❹	quiz_score is REQUIRED
3	unique_ name	Unique	name	name must be unique

In Example 6.4, there are three constraints that are created. The first constraint, *req_quiz_score*, is the NOT NULL constraint that is created for the *quiz_score* variable ❶. This constraint will confirm that the *quiz_score* variable does not contain missing values. The second constraint is *ok_class*, which is the CHECK constraint ❷. *Ok_class* will check the values of the *class* variable to make sure that they are in the list specified in the constraint. The last constraint is the *unique_name* constraint, which is the UNIQUE constraint ❸. The *unique_name* constraint will confirm the *name* variable contains only unique values. **Once the constraints are established, SAS will not allow the entry of invalid data into these variables.**

The output shows the listing created by the CONTENTS procedure. The procedure will show the constraints, types associated with the constraint, variables, any WHERE clauses associated with the constraint, and the user message ❹.

6.2.4 *Tracking Data Updates with Audit Trails*

Syntax:
PROC DATASETS;
  AUDIT <SAS name>;
  INITIATE;
  USER_VAR <variable1... varaiblen>;
  <suspend | resume | terminate>;
QUIT;

Audit trails are useful for tracking changes to your data. An audit trail is created by the DATASETS procedure. In the DATASETS procedure, the AUDIT statement is used to specify the data set that will be audited. An audit trail file is created with the same name as the data set in the same SAS data library but as a member type of AUDIT. The audit is then initiated by using the statement INITIATE. In addition to the INITIATE statement, there are three other statements to control the audit file. The SUSPEND statement will suspend event logging to the audit file, but does not delete the audit file; the RESUME statement will resume event logging to the audit file, if it was suspended; and the TERMINATE statement will terminate event logging and delete the audit file. The USER_VAR statement allows you to create your own audit variables in the data set that are in addition to the six audit variables that SAS automatically creates. If you add records to the data set, you are required to enter values for the user audit variables. The six automatic variables are listed in Table 6.1, along with the codes for the _ATOPCODE_ in Table 6.2. **Note that in order for SAS's audit trail feature to become effective, you must modify the data set instead of recreating the data set.** You can use the MODIFY statement in the DATA step or the SQL procedure to insert data in the data set instead of specifying the MERGE, SET, or UPDATE statements within the DATA step to recreate the data set.

Table 6.1 Automatic Audit Trail Variables

Variable	Description
ATDATETIME	Stores the date and time of modification.
ATUSERID	Stores the logon userid associated with the modification
ATOBSNO	Stores the observation number affected by the modification
ATRETURNCODE	Stores the event return code
ATMESSAGE	Stores the SAS log message at the time of the modification
ATOPCODE	Stores a code describing the type of operation (listed below)

Table 6.2 _ATOPCODE_ Values

Code	Event
DA	Data added record image
DO	Data deleted record image
DR	Before-update record image
DW	After-update record image
EA	Observation add failed
ED	Observation delete failed
EU	Observation update failed

Example 6.5 Tracking data updates with audit trails.

```
libname mylib 'c:\myproject1';

data mylib.tests;
   input name $ class $ quiz_score ;
   cards;
   Tim math 9
   Steve history 7
   James science 7
   Sally math 10
   Michelle science 7
   Adam history 5
   Doug math 8
```

```
   Lynn history 7
   Dave science 6
;
run;

Proc datasets lib=mylib;
   audit tests;
   initiate; ❶
   user_var chgreason $50 label = 'Reason for change';
quit;

proc sql;
 insert into mylib.tests
 set name = 'Tim',
 class = 'reading', ❷
 quiz_score = 9,
 chgreason = 'update';
quit;

Proc Datasets lib=mylib;
   audit tests;
   suspend;
quit;

proc print data=mylib.tests (type=audit);
run;
```

Output

Obs	name	class	quiz_score	chgreason	_ATDATETIME_	_ATOBSNO_	_ATRETURNCODE_	_ATUSERID_	_ATOPCODE_	_ATMESSAGE_
1	Tim	reading	9	update	15AUG2004: 20:50:03	10	.	sunil gupta	DA	❸

In Example 6.5, once the audit trail feature is initiated ❶, SAS can track changes to the data set if updates are made. Notice that the record inserted with the SQL procedure ❷ was documented as data added DA in the _ATOPCODE_ variable and the date and time of data entry ❸ was entered in the _ATDATETIME_ variable.

6.2.5 Backing Up Data with Generation Data Sets

Syntax:
Genmax data set option:
DATA <data set name> (genmax=number of generations);
RUN;

Gennum data set option:
PROC PRINT DATA=data set name (gennum=generation number);
RUN;

Data sets can be backed up with generation data sets by using the GENMAX= data set option. The GENMAX= data set option is specified in parentheses after the data set name in the DATA step. The value specified after the GENMAX= data set option is the maximum number of generation data sets that will be created. The value can be from 0 to 100; a value of 0 is the default and means that no generation data sets will be created. The first time the data set is replaced, SAS keeps the replaced version and appends a four-character version number to its member name, which includes "#" and a three-digit number. Since a four-character version number is appended to the data set name, the data set name can only be 28 characters.

After recreating the data sets specified with the GENMAX= data set option, you can use the GENNUM= data set option to access a specific version. The value specified in the GENNUM= data set option is the number of the historical version data set. If GENNUM=1 was specified then the most current data set is selected.

Example 6.6   Backing up data with generation data sets.

```
libname mylib 'c:\myproject1';

data mylib.tests;
   input name $ class $ quiz_score ;
   cards;
   Tim math 9
   Steve history 7
   James science 7
```

```
    Sally math 10
    Michelle science 7
    Adam history 5
    Doug math 8
    Lynn history 7
    Dave science 6

  ;
  run;


  data mylib.tests (genmax=3); ❶  /* first version */
     set mylib.tests;                /* tests#001   */
  run;


  data mylib.tests; ❷              /* second version */
     set mylib.tests;              /* tests#002    */
     where class='math';
  run;


  data mylib.tests; ❸              /* third version */
     set mylib.tests;              /* tests#003    */
     where name='Tim';
  run;
                      ❹             /* Alternative reference   */
  proc print data=mylib.tests (gennum=2);/* proc print */
  /* data=mylib.tests#002;                              */
     run;                         /* run; */
```

Output

| | | ❺ | quiz_ |
Obs	name	class	score
1	Tim	math	9
2	Sally	math	10
3	Doug	math	8

In Example 6.6, GENMAX=3 instructs SAS to create three versions of the data set ❶. The three data sets that are created are the current version and two historical data sets. The oldest data set is tests#001 which was created with the first DATA step. The next oldest data set is tests#002 which is created with the subsetting of class='math' ❷. The third data set, tests, is the most current and is created by the last DATA step where the data is subset by name='Tim' ❸. Since GENNUM=2 is specified in the PRINT procedure, the output is from the second version data set where the data was subset by class='math' ❹. This can be seen in the output window ❺. If the data set is created again, then the first most recent version, tests#003, is moved to the second most recent version, tests#002, and is replaced by the current version. The oldest version, tests#001, is deleted and replaced by the original second most recent version, tests#002.

6.3   Version 9.1 Enhancements

6.3.1   Taking Advantage of Multiprocessing Capabilities

Syntax:
OPTIONS THREADS;

Specifying the THREADS option in the system options statement instructs SAS to use threaded processing if it is available in the operating system. This option is available at the system level as well as at the SAS procedure level for selected procedures. This option is useful for reducing the processing time of your SAS programs. See SAS online DOC 9 for more information.

6.3.2   HTML File with Custom Text

Syntax:
PROC FREQ data=dataset name;
  Tables variable1*variable2/ contents='contents text';
Run;

For selected SAS procedures, such as PROC FREQ, you can now directly specify the text to be placed in the HTML contents file. The syntax for specifying text to be placed in the HTML contents file is the CONTENTS= option followed by the text to be displayed. Note that the custom text will not appear in the RTF or PDF files because this option is only applicable to the HTML contents file. See Section 4.7 in Chapter 4 for more information on HTML files and ODS.

Example 6.7   Creating an HTML file with custom text.
```
data mylib.tests;
   input name $ class $ quiz_score ;
   cards;
```

```
    Tim math 9
    Steve history 7
    James science 7
    Sally math 10
    Michelle science 7
    Adam history 5
    Doug math 8
    Lynn history 7
    Dave science 6
;
run;

ods html body = 'c:\freqs_body.html'
         contents = 'c:\freqs_contents.html'
         frame = 'c:\freqs_out.html'
;

proc freq data=mylib.tests;
   table class*quiz_score / Contents='My Contents';
run;

ods html close;
```

Output

The SAS System

The FREQ Procedure

Frequency Percent Row Pct Col Pct			Table of class by quiz_score					
					quiz_score			
	class	5	6	7	8	9	10	Total
	history	1 11.11 33.33 100.00	0 0.00 0.00 0.00	2 22.22 66.67 50.00	0 0.00 0.00 0.00	0 0.00 0.00 0.00	0 0.00 0.00 0.00	3 33.33
	math	0 0.00 0.00 0.00	0 0.00 0.00 0.00	0 0.00 0.00 0.00	1 11.11 33.33 100.00	1 11.11 33.33 100.00	1 11.11 33.33 100.00	3 33.33
	science	0 0.00 0.00 0.00	1 11.11 33.33 100.00	2 22.22 66.67 50.00	0 0.00 0.00 0.00	0 0.00 0.00 0.00	0 0.00 0.00 0.00	3 33.33
	Total	1 11.11	1 11.11	4 44.44	1 11.11	1 11.11	1 11.11	9 100.00

As can be seen in Example 6.7, the custom text "My Contents" is placed as the hyperlink in the table of contents section of the HTML file.

6.3.3 DUPOUT= Option to Save Duplicate Records

Syntax:
PROC SORT DATA=input data set <options> DUPOUT=duplicate records data set

                                                          OUT=output data set;
  BY <variable1 ... variablen>;
RUN;

For the first time, SAS Version 9.1 makes it easy to save the records that would normally not be saved when applying a NODUPKEY or NODUP option in the SORT procedure. Any record not saved to the output data set from the original data set is saved to the duplicate records data set. This becomes very helpful for reviewing only the duplicate records and not the unique records.

Example 6.8   Using the DUPOUT= option to save duplicate records.

```
data tests;
   input name $ class $ test_score ;
   cards;
   Tim math 9
   Tim history 7
   Tim science 7
   Sally math 10
   Sally science 7
   Sally history 5
   John math 8
   John history 7
   John science 6
;
run;
```
                              ❶   ❷
```
proc sort data=tests nodupkey dupout = dup_tests
out=tests_sort;
   by name;
run;

title 'All Data';
```

```
proc print data=tests;
run;

title 'Non-Duplicate Key Records: name';
proc print data=tests_sort;
run;

title 'Deleted Duplicate Records';
proc print data=dup_tests;
run;
```

Output

All Data

Obs	name	class	test_ score
1	Tim	math	9
2	Tim	history	7
3	Tim	science	7
4	Sally	math	10
5	Sally	science	7
6	Sally	history	5
7	John	math	8
8	John	history	7
9	John	science	6

Non-Duplicate Key Records: name ❸

Obs	name	class	test_ score
1	John	math	8
2	Sally	math	10
3	Tim	math	9

Deleted Duplicate Records   ❹

Obs	name	class	test_ score
1	John	history	7
2	John	science	6

```
3    Sally    science    7
4    Sally    history    5
5    Tim      history    7
6    Tim      science    7
```

In Example 6.8, the DUPOUT= option ❷ is useful for saving the duplicate records in a separate data set. Note that this process also requires you to specify the NODUPKEY or NODUP option ❶ to remove duplicate records in the final data set. The data set specified in the OUT= option does not change. Only the first record with a unique *name* key variable is saved in the *tests_sort* data set ❸. The new duplicate records data set, *dup_tests*, contains all of the other records that were not saved in the *tests_sort* data set ❹. These records contain the duplicate name key variable.

6.3.4 Changes to the SQL Procedure

Although the syntax of the SQL procedure has not changed, there are two enhancements to the actual procedure. The SQL procedure can now reference up to 32 views and tables. In addition there are more dictionary tables and new columns in existing tables. The INTO clause now also supports leading zeros as macro names when creating macro variables.

Example 6.9 Using the SQL procedure to create macro variables.

```
data drug;
   input gender $ race $ drug $ dose @@;
   cards;
M White Active 5      M White Active 5
F White Placebo 6     M Nonwhite Active 6
M Nonwhite Placebo 4 F White Active 7
M Nonwhite Active 5   M Nonwhite Placebo 7
F White Active 6      F White Active 4
;
run;


proc sql;
select gender into :n01 - :n10 from drug; ❶
quit;
%put n01=  &n01   n02=  &n02   n10=  &n10;
```

Log

```
102   proc sql;
103   select gender into :n01 - :n10 from drug;
104   quit;
NOTE: PROCEDURE SQL used (Total process time):
      real time              0.01 seconds
      cpu time               0.02 seconds



105   %put n01=  &n01   n02=  &n02   n10=   &n10;
n01=   M   n02=   M   n10=   F
```

In Example 6.9, SAS creates ten macro variables that are named *n01* through *n10* ❶. Each macro variable contains the value of the *gender* variable from each record of the data set.

6.3.5 New Character and Numeric Functions

There are over 200 new functions in Version 9.1. New character and numeric functions allow greater flexibility in meeting programming requirements. This section discusses some of the functions, as listed in Table 6.3, that may be commonly used.

Table 6.3   Selected Version 9.1 Functions

Function	Description
ANYALNUM()	Returns the first position of alphanumeric character
ANYALPHA()	Returns the first position of alphabetic character
CAT(), CATT(), CATS(), CATX()	The family of CAT functions used to help reduce the complexity of concatenating strings
COUNT()	Used to help replace INDEX() and SUBSTR() to count number of occurrences of a text within a string
COMPARE()	Returns the leftmost character position when two strings differ or the value 0 if no difference exists
MEDIAN()	Computes the median value from a list of non-missing values
PCTL()	Computes percentiles
SYMPUTX()	Automatically converts numeric to character values and strips leading and trailing blanks

6.3.6 ANYALNUM() Function

Syntax:
<Variable> = ANYALNUM(string,<starting position>);

The ANYALNUM( ) function returns the position of the first alphanumeric character in a string. **If the starting position is not specified, SAS will start searching at the beginning of the string.** The syntax of the ANYALNUM() function is the function name followed by the string that will be scanned and then the optional starting position, all in parentheses.

Example 6.10   Using the ANYALNUM( ) function to identify any alpha numeric character.

```
data _null_;
   string= 'abc 123 +=/';
   alpha=anyalnum(string);
   numeric=anyalnum(string,5);
   other = anyalnum(string,9);
   put alpha= numeric= other=;
run;
```

Log (Partial)

```
74    data _null_;
75       string= 'abc 123 +=/';
76       alpha=anyalnum(string);
77       numeric=anyalnum(string,5);
78       other = anyalnum(string,9);
79       put alpha= numeric= other=;
80    run;
          ❶          ❷          ❸
alpha=1 numeric=5 other=0
NOTE: DATA statement used (Total process time):
         real time           0.07 seconds
         cpu time            0.03 seconds
```

In Example 6.10, the first variable, *alpha,* is set to 1, which identifies the first position of the alphanumeric character in the string 'abc 123+ =/;❶. Since the starting position was not specified, SAS started searching at the beginning of the string. The second variable, *numeric,* is set to 5 since it is the first alphanumeric character starting at the fifth position in the string ❷. The last variable, *other,* is set to zero.❸ The starting position specified is 9. There are no alphanumeric characters after the ninth position so the variable is set to zero.

6.3.7 ANYALPHA() Function

Syntax:
<Variable>=ANYALPHA(string,<starting position>);

The ANYALPHA() function returns the position of the first character in a string. **If the starting position is not specified, SAS will start searching at the beginning of the string.** The syntax of the ANYALPHA() function is the function name followed by the string that will be scanned and then the optional starting position, all in parenthesis.

Example 6.11   Using the ANYALPHA() function to identify characters.

```
data _null_;
   string= 'abc 123 +=/';
   alpha=anyalpha(string);
   numeric=anyalpha(string,5);
   other = anyalpha(string,9);
   put alpha= numeric= other=;
run;
```

Log (Partial)

```
81        data _null_;
82        string= 'abc 123 +=/';
83        alpha=anyalpha(string);
84        numeric=anyalpha(string,5);
85        other = anyalpha(string,9);
86        put alpha= numeric= other=;
87     run;
```

         ❶   ❷   ❸
alpha=1 numeric=0 other=0
NOTE: DATA statement used (Total process time):
       real time   0.04 seconds
       cpu time   0.04 seconds

In Example 6.11, the first variable, *alpha,* is set to 1, which is the location of the first position of the alphabetic character in the string 'abc 123+=\'❶. Since there was no starting position specified, SAS started searching from the beginning of the string. The second variable, *numeric,* has the starting position set to 5. SAS will start searching the string at the fifth position; since there are no alphabetic characters after the fifth position, the variable is set to zero ❷. The last variable, *other,* is also set to zero ❸ since there are no alphabetic characters after the ninth position, which is the specified starting position.

6.3.8 CAT(), CATT(), CATS(), and CATX() Functions

Syntax:
<Variable>=CAT(string1,...,stringn);
<Variable>=CATT(string1,...,stringn);
<Variable>=CATS(string1,...,stringn);
<Variable>=CATX('',string1,...,stringn);

Table 6.4   V 9.1 Cat Function Code in version 8.2

CAT(X1,X2,X3)	**X1 | | X2 | | X3**
CATT(X1,X2,X3)	**TRIM(X1) | | TRIM(X2) | | TRIM(X3)**
CATS(X1,X2,X3)	**TRIM(LEFT(X1)) | | TRIM(LEFT(X2)) | | TRIM(LEFT (X3))**
CATX(SP, X1, X2,X3)	**TRIM(LEFT(X1)) | | SP | | TRIM(LEFT(X2)) | | SP | | TRIM(LEFT(X3))**

In Table 6.4, the CAT(), CATS(), CATT(), and CATX() functions replace the combinations of the concatenation operator (| |) and the TRIM() and LEFT() functions. The CAT(), CATS(), CATT(), and CATX() functions are faster and easier than specifying the TRIM() and LEFT() functions. The CAT() function will just concatenate the strings together while the CATT() and CATS() functions will either trim the strings before the concatenation or trim and left justify the strings before concatenation, respectively. The CATX() function will trim and left justify the strings and will allow you to specify the separator.

Example 6.12   Using the CATX() function to concatenate strings.

```
data test;
   length a b $10;
   a = ' Good ';
   b = 'Afternoon';
   v9func= CATX('',a,b); ❶
   oldway=trim(left(a))||' '||trim(left(b));
run;

proc print data=test;
run;
```

Output

Obs	a	b	v9func ❷	oldway
1	Good	Afternoon	Good Afternoon	Good Afternoon

In Example 6.12, the CATX() function will first left justify and trim the strings and then concatenate the strings using the specified separator, which is a space ❶. For the variable *a* the leading space will be removed before the concatenation. The *oldway* variable shows the same results using the traditional functions TRIM() and LEFT(). ❷

6.3.9 COUNT() Function

Syntax:
<Variable>=COUNT(string,substring,<modifiers>);

Modifiers
I = Ignores the case when counting
T = Trims blanks from the string

The COUNT() function helps reduce the complexity of using the INDEX() and SUBSTR() functions to count the number of occurrences of a text within a string. The syntax of the COUNT() function is the function name followed by the string that will be searched and then the search substring, all in parentheses. There are also a couple of modifiers that can be specified after the substring. The I modifier instructs SAS to ignore the case during the count. If this modifier is not specified then SAS will only count the character substrings that match the substring. The T modifier will trim trailing blanks from the string and substring.

Example 6.13   Using the COUNT() function to tabulate totals.

```
data test;
  A='Good Afternoon.';
  X_00=count(a,'oo'); ❶
run;

proc print data=test;
run;
```

Output

```
        Obs            A            X_00

         1      Good Afternoon.      2
```

In Example 6.13, the COUNT() function is used to create the variable *x_00*, which is the count of the number of occurrences of the string "oo" in the variable *a* ❶. In this example, all of the combinations of "oo" were in lowercase so the I modifier was not needed. Also, there were no trailing spaces, so the T modifier was also not needed.

6.3.10 COMPARE() Function

Syntax:
<Variable>= COMPARE(string1, string2, modifiers);

Modifiers
i or I = Ignores the case when comparing
l or L = Removes leading blanks before comparing

The COMPARE() function returns the leftmost character position when two strings differ or the value 0 if no difference exists. The sign of the result is negative if *string1* proceeds *string2* in a sort sequence and positive if *string1* follows *string2* in a sort sequence. The syntax of the COMPARE() function is the function name followed by the strings that will be compared and then any modifiers. The strings can be specified as a variable or as a string enclosed in quotes. The i or I modifier is used when you want to ignore the case when comparing. The l or L modifier is used to remove leading blanks before the compare is made.

Example 6.14   Using the COMPARE() function to compare strings.

```
data test;
   input string1 $8. string2 $8. modifiers $8.;
   Result=compare(string1,string2,modifiers);
   datalines;
123        abc
abc        abx
aBc        AbC        i
   abc     abc        l
;
run;

proc print data=test;
run;
```

Output

Obs	string1	string2	modifiers	Result	
1	123	abc		-1	❶
2	abc	abx		-3	❷
3	aBc	AbC	i	0	❸
4	abc	abc	l	0	❹

In Example 6.14, the first observation compares the string "123" with "abc". The first character is different between the two strings and string "123" is before "abc" in a sort, so the result is -1.❶ In the second observation the difference is in the third character and again the first string is before the second in a sort, so the result is -3.❷ In the third observation the i modifier is used, which will cause the COMPARE() function to ignore the case when the strings are compared. So, in this observation the result will be 0 since the strings are the same when the case is ignored.❸ In the last observation the l modifier is used, which will cause the COMPARE() function to remove the leading blanks from the first string before the compare. With the leading blanks removed the strings are the same, so the result will be 0. ❹

6.3.11   MEDIAN() Function

Syntax:
<Variable>=MEDIAN(variable1, … , variablen);

The MEDIAN() function computes the median value of a list of non-missing values. The syntax is the function name followed by the variables or values that the median is calculated for. The values or the variables must be non-missing.

Example 6.15   Using the MEDIAN() function to calculate median values.

```
data test;
   N1=1;
   N2=2;
   N3=3;
   N4=4;
   Med=median(n1, n2, n3, n4);
run;

proc print data=test;
run;
```

Output

Obs	N1	N2	N3	N4	Med
1	1	2	3	4	2.5

In Example 6.15, the *med* variable is set to 2.5, which is the median of the variables *n1*, *n2*, *n3*, and *n4*.

6.3.12   PCTL() Function

Syntax:
<Variable>=PCTL(percentile,variable1, ... , variablen);

The PCTL() function computes the specified percentile of a list of non-missing values. The syntax is the function name followed by the percentile and then the variables or values that the percentile is calculated for. The values or the variables must be non-missing.

Example 6.16   Using the PCTL() function to calculate percentiles.

```
data test;
   N1=1;
   N2=2;
   N3=3;
   N4=4;
   pctvl=pctl(25,n1, n2, n3, n4);
run;

proc print data=test;
run;
```

Output

Obs	N1	N2	N3	N4	pctvl
1	1	2	3	4	1.5

In Example 6.16, the PCTL() function is used to create the *pctvl* variable, which is the 25th percentile of the variables *n1* through *n4*.

6.3.13   SYMPUTX() Function

Syntax:
CALL SYMPUTX(macro variable, numeric value);

The SYMPUTX() function automatically converts numeric to character values and strips leading and trailing blanks, thus creating a character macro variable from a numeric value or variable. The syntax of the function is the keyword CALL followed by the function name. In parentheses after the function name is the macro variable that you are creating followed by the numeric string or variable.

Example 6.17 Using the SYMPUTX( ) function to create macro variables.

```
data _null_;
  call symputx('v9way',99.9); ❶
  call symput('oldway',trim(left(put(99.9,4.1)))); ❷
run;

%put &v9way &oldway;
```

Log

```
1    data _null_;
2      Call symputx('v9way' , 99.9);
3      Call symput('oldway' , trim(left(put(99.9,4.1))));
4      run;

NOTE: DATA statement used (Total process time):
      real time             0.29 seconds
      cpu time              0.09 seconds


5
6    %put &v9way &oldway;
99.9 99.9 ❸
```

In Example 6.17, the SYMPUTX() function replaces ❶ the TRIM(), LEFT(), and PUT() functions that were used in addition to the SYMPUT() function in previous SAS versions. ❷ The SAS log shows that both methods will give the same result in the conversion of the numeric value into a character macro variable. ❸

See the following SAS papers and books for more information on version 8.2 and 9.0: *Getting Familiar with SAS® Version 8.2 and 9.0 Enhancements, What's New in Versions 7 and 8 for SAS Files, What's New in the Output Delivery System, Version 9.0, Several Ways to Tune a SORT, Version 6 and Version 7: A Peaceful Co-Existence, Making the Most of Version 9 Features and Power SAS*, and *An Introduction to ODS for Statistical Graphs in SAS 9.1.*

Chapter 6.   Version 8.2 and Version 9.0 Enhancements —Chapter Summary

Concatenating Libraries and Catalogs

LIBNAME *libref<engine>* (*library-specification-1 <. . . library-specification-n>*) *< options >*;
CATNAME*<libref.> catref* (*libref-1.catalog-1 ...libref-n.catalog-n*);

Preventing Data Errors with Integrity Constraints

```
Proc datasets;
   Modify <data set name>;
   Integrity constraint create
   <constraint name> = <constraint>     /*  not null, unique, check  */
   message = 'message string';
quit;
```

Tracking Data Updates with Audit Trails

```
Proc datasets;
   Audit <SAS name>;
   Initiate;
   User_var <variable1... varaiblen>;
   <suspend | resume | terminate>;
quit;
```

Backing Up Data with Generation Data Sets

```
Genmax data set option:
Data <data set name> (genmax=number of generations);
Run;
```

```
Gennum data set option:
Proc print data=data set name (gennum=generation number);
Run;
```

Selected Version 9.1 Functions

Function	Description
ANYALNUM( )	Returns the first position of alphanumeric character
ANYALPHA( )	Returns the first position of alphabetic character
CAT( ), CATT( ), CATS( ), CATX( )	The family of CAT functions used to help reduce the complexity of concatenating strings
COUNT( )	Used to help replace INDEX( ) and SUBSTR( ) to count number of occurrences of a text within a string
COMPARE( )	Returns the leftmost character position when two strings differ or the value 0 if no difference exists
MEDIAN( )	Computes the median value from a list of non-missing values
PCTL( )	Computes percentiles
SYMPUTX( )	Automatically converts numeric to character values and strips leading and trailing blanks

Chapter 6.   Version 8.2 and Version 9.1 Enhancements —Chapter Questions

Question 1: Which system option can be used to run a Version 6.12 SAS program from within SAS Version 8 or higher?

Question 2: Name at least two types of integrity constraints available to improve the quality of data entering data sets.

Question 3: What is the maximum number of generation data sets that SAS can create when backing up data using the GENMAX= data set option.

Question 4: What is the alternative method for referencing the following generation data set: tests (gennum=2)?

Question 5: In Version 9.1, how many views or tables can be referenced in the SQL procedure?

Glossary

ARRAYS A convenient way of temporarily identifying a group of variables.

ATTRIB Used to specify label, format, and length of variables.

AUDIT TRAILS SAS file created by the SAS/BASE engine with the same libref and member name as the data file, and a data set type of AUDIT.

CARDS Used to signal that data starts on the next line within the DATA step.

COLUMN INPUT Used to read data in specified locations in the data file.

COLUMN POINTER CONTROL @ Used to place the pointer at a specific column. This is usually placed just before the variable name.

COMPILE PHASE First time SAS passes through code in which it compiles the SAS statements, checks the syntax, creates the input buffer and PDV, and creates descriptor portion of data set.

COMPRESS( ) Removes specific characters from a character string.

DATA SET OPTIONS Specify actions that apply only to the SAS data set with which they appear.

DATA _NULL_ Keyword to instruct SAS not to create data set.

DATALINES Used to signal that data starts on the next line within the DATA step.

DATES/TIMES Special types of nonstandard data.

DO TO Executes statements in a DO loop repetitively until the last specified number or condition is reached. Performs top evaluation.

DO UNTIL Executes statements in a DO loop repetitively until a condition is true. Performs bottom evaluation.

DO WHILE Executes statements repetitively while a condition is true. Performs top evaluation.

DOUBLE TRAILING @@ CONTROL The INPUT statement for the next iteration of the DATA step continues to read the same record.

DROP Excludes variables from the output SAS data set.

EXECUTE PHASE Second pass through of the code in which the SAS statements are executed.

FILENAME Associates an SAS fileref with an external file or an output device.

FOOTNOTES Used to display report footnotes.

FORMAT Associates formats with variables to specify the display of data values.

FORMAT MODIFIER Can be used with the LIST INPUT statement: colon (:), ampersand (&), tilde (~).

FORMATTED INPUT Used to read dates and other nonstandard data.

GENERATION DATA SETS Historical copies of a SAS data set.

IF Continues processing only those observations that meet the condition.

IF–THEN–ELSE Executes a SAS statement for observations that meet the specific conditions.

INDEX( ) Searches a character expression for a string of characters.

INFILE Identifies an external file to read with an INPUT statement.

INFORMAT Associates informats with variables to specify how data values are read.

INPUT( ) Returns the value produced when a SAS expression that uses a specified informat expression is read.

INPUT BUFFER Place where SAS stores raw data.

INTCK( ) Returns the integer number of time intervals in a given time span.

INTEGRITY CONSTRAINTS Set of data validation rules that you can specify to restrict the data values accepted into a SAS data set.

KEEP Includes variables in the output SAS data set.

LABEL Assigns descriptive labels to variables.

LEFT( ) Left-aligns a SAS character expression.

LENGTH Specifies the number of bytes for storing variables.

LIBREF Created by the LIBNAME statement and used to reference data sets.

LINE POINTER CONTROL / #N Specifies which input record to read.

LIST INPUT Used to read standard data in order listed in data file.

MEAN( ) Returns the arithmetic mean (average).

MERGE Joins observations from two or more SAS data sets into single observations. Called type1 merge; one-to-one.

MERGE WITH BY Joins observations from two or more SAS data sets into a single observation in a new data set according to the values of BY variables. Called type2 merge; match-merge.

MIN( ) Returns the smallest value.

MONTH( ) Returns the month from a SAS date value.

N( ) Returns the number of nonmissing values.

NAMED INPUT Used to read data containing a specified text.

NONSTANDARD DATA Numeric data that contains special character or data that are dates or times.

ODS Output delivery system. Used to create files such as HTML, RTF, PDF, and Excel.

PROC APPEND Adds observations from one data set to the end of another data set.

PDV Program data vector. Place where SAS stores data values and variable attributes.

PROC CONTENTS Displays the variable attributes along with number of observations and variable names.

PROC COPY Copies members from a library to another library.

PROC DATASETS Enables you to list, copy, rename and delete SAS files along with append data sets.

PROC EXPORT Reads data from a SAS data set and writes it to an external data source.

PROC FORMAT Define user-defined formats.

PROC FREQ Produces one-way to *n*-way frequency and cross-tabulation tables.

PROC IMPORT Reads data from an external data source and writes it to a SAS data set.

PROC MEANS Produces simple and univariate descriptive statistics.

PROC PRINT Print the observations from a data set.

PROC REPORT Prints and summarizes data sets in table format.

PROC SORT Sorts observations in a data set.

PROC SQL Enables you to create and query data sets.

PROC TABULATE Displays descriptive statistics in tabular form.

PUT Writes lines to the SAS log, to the SAS procedure output file, or to an external file that is specified in the most recent FILE statement.

PUT( ) Returns a character value using a specified format.

RENAME Specifies new names for variables in the output SAS data set.

RETAIN Causes a variable that is created by an INPUT or assignment statement to retain its value from one iteration of the DATA step to the next.

RIGHT( ) Right-aligns a character expression.

SCAN( ) Selects a given word from a character expression.

SELECT–WHEN Executes one of several statements or groups of statements.

SET Reads an observation from one or more SAS data sets.

SET WITH BY Reads an observation from one or more SAS data sets and orders the observations by the BY variables.

SQL: COALESCE Select an observation for a variable from the first data set where the variable is non-missing.

SQL: INNER JOIN Returns a result table for all of the rows in a table that have one or more matching rows in another table.

SQL: OUTER FULL JOIN Similar to the RIGHT and LEFT JOIN except that all of the matching and nonmatching observations are kept.

SQL: OUTER LEFT JOIN Similar to an INNER JOIN except that all of the rows from the first data set are included instead of just the matching rows in the new data set.

SQL: OUTER RIGHT JOIN Similar to an INNER JOIN except that all of the rows from the second data set are included instead of just the matching rows in the new data set.

STANDARD DATA Data that contains only characters, numbers, decimal points, and a plus or minus sign.

STYLE ODS uses styles to display the report.

SUBSTR( ) Extracts a substring from an argument.

SUM( ) Returns the sum of the nonmissing arguments.

SYSTEM OPTIONS Options that control the way that SAS performs operations such as SAS system initialization, hardware and software interfacing, and the input, processing, and output of jobs and SAS files.

THREADS System option in Version 9.1 to improve performace.

TITLES Used to display report titles.

TODAY( ) Returns the current date as an SAS date value.

TRAILING @ CONTROL Allows the next INPUT statement to read from the same record.

TRIM( ) Removes trailing blanks from character expressions and returns one blank if the expression is missing.

UPCASE( ) Converts all letters in an argument to uppercase.

USER-DEFINED FORMAT Formats created by user.

WHERE Selects observations from SAS data sets that meet a particular condition.

Tables and Figures

References

Articles, Papers and Books

Alonso, Richard, *Manipulating SAS Datasets: An Introduction.*

Aster, Rick, and Seidman, Rhena, *Professional SAS Programming Secrets,* New York, NY: Windcrest/McGraw-Hill, 1991.

Beatrous Steve, and Holman, James, *Version 6 and Version 7: A Peaceful Co-Existence,* Training and User Support Services.

Bhat, Gajanan, and Suligavi, Raj, "Merging Tables in DATA Step vs. PROC SQL: Convenience and Efficiency Issues," Coder's Corner, SUGI 26.

Burlew, Michele M., *Debugging SAS Programs: A Handbook of Tools and Techniques,* Cary, NC: SAS Institute, Inc., 2001.

Bruns Dan, Pass, Ray, and Eaton, Alan, "Battle of the Titans: REPORT vs. TABULATE," SUGI 27, Data Presentation.

Bruns, Dan, "The Power and Simplicity of the Tabulate Procedure," SUGI 25, Hands-On Workshop.

Bruns, Dan, "The Utter 'Simplicity?' of the Tabulate Procedure—The Final Chapter?" SUGI 29, Tutorials.

Carey, Helen and Carey, Ginger, "180 Terrific Ideas for Using and Learning the SAS System," SUGI 18.

Chapman, David D., "Using Proc Report To Produce Tables With Cumulative Totals and Row Differences," SUGI 27, Data Presentation.

Cody, Ron, *Cody's Data Cleaning Techniques Using SAS Software,* Cary, NC: SAS Institute, Inc., 1999.

Cody, Ron, *The SAS Workbook,* Cary, NC: SAS Institute, Inc., 1996.

Cody, Ron, and Pass, Ray, *SAS Programming by Example,* Cary, NC: SAS Institute, Inc., 1995.

Coleman, Ron, "The Building Blocks of Proc Tabulate," SUGI 23, Application Development.

Delwiche, Lora, and Slaughter, Susan, "Errors, Warnings, and Notes (Oh My), A Practical Guide to Debugging SAS Programs," SUGI 22.

Delwiche, Lora, and Slaughter, Susan, *The Little SAS Book: A Primer, Revised Second Edition,* Cary, NC: SAS Institute, Inc. 2002.

Dickson, Alan, *Blind Dates and Other Lapses of Reason: Handling Dates in SAS,* NESUG, Baltimore, MD, 1992.

Dilorio, Frank C., *SAS Applications Programming: A Gentle Introduction,* Belmont, California, Duxbury Press, 1991.

Dilorio, Frank C., *The Elements of SAS Programming Style.*

Dilorio, Frank C., "The SAS Debugging Primer," SUGI 26.

Doty, Ann, *Programming with SAS Functions, with a Special Emphasis on SAS Dates and Times.*

Ewing, Daphne, and Pass, Ray, *So Now You're Using PROC REPORT. Is it Pretty and Automated?*

Fecht, Marje, "Making the Most of Version 9 Features," SUGI 28.

First, Steven, "Developing Easy Maintained SAS Code," SUGI 19.

Gill, Paul, *The Next Step: Integrating the Software Life Cycle with SAS Programming,* Cary, NC: SAS Institute, Inc., 1997.

Gilsen, Bruce, *SAS Program Efficiency for Beginners,* SUGI 22, Beginning Tutorials.

Gupta, Sunil, "Getting Familiar with SAS® Version 8.2 and 9.0 Enhancements," Proceedings of the Eleventh Annual Western Users of SAS Software, San Francisco, CA.

Gupta, Sunil, *Quick Results with the Output Delivery System,* Cary, NC: SAS Institute, Inc., 2003.

Gupta, Sunil, "SAS®'s ODS Technology for Today's Decision Makers," Proceedings of the Eleventh Annual Western Users of SAS Software, San Francisco, CA.

Gupta, Sunil, "Utilizing Clinical SAS Report Templates with ODS," Proceedings of the Tenth Annual Western Users of SAS Software, San Diego, CA.

Gupta, Sunil, "Using Styles and Templates to Customize SAS ODS Output," SUGI 29.

Gupta, Sunil, and Shipp, Charles E., "Utilization of SAS Programs in the Business Environment," Poster Presentation, SUGI 25.

Haase, Martin, "PROC Report: A Guided Tour," WUSS 95, Data Presentation.

Hall, Robert G. and Mopsick, Judith H., "Getting the Most from the SAS Log," SUGI 19.

Harris, Michael C., *Software Validation for the Rest of Us,* Western Users of SAS Software, 1998.

Hamilton, Jack, "The Problem with Noduplicates," SUGI 25, Posters.

Hardy, Jean, *Proc Report Batch Language: Tips And Techniques.*

Haworth, Lauren, *Anyone Can Learn PROC TABULATE, v2.0* SUGI 25, Beginning Tutorials.

Haworth, Lauren, *HTML for the SAS Programmer* SUGI 25, Internet and Intranets on the Web.

Howard, Neil, "Data Step Internals: Compile and Execute," SUGI 28.

Howard, Neil, "Discovering the FUN in SAS Functions," SUGI 19.

Howard, Neil, "It's Not a Bug, It's a Feature!!" SUGI 21.

Jaffe, Jay, *Mastering the SAS System, Second Edition,* New York, NY: Van Nostrand Reinhold, 1994.

Jensen, Edward, "Tapping the Power of Proc Tabulate for Analysis Reports Clients Understand."

Johnston, Susan E., *Using the SQL Procedure in SAS Programs.*

Karp, Andrew H., "Steps to Success with Proc Means," Tutorials, SUGI 29.

Knapp, Peter, "Debugging 101."

Lafler, Kirk Paul, *Diving into SAS Software with the SQL Procedure,* SUGI 23.

Lafler, Kirk Paul, *Frame Your View of Data with the SQL Procedure.*

Lafler, Kirk Paul, *Power SAS: A Survival Guide,* New York, NY: Apress, 2002.

Lafler, Kirk Paul, *Proc SQL: Beyond the Basics Using SAS,* Cary, NC: SAS Institute, Inc. 2004.

Lafler, Kirk Paul, *Querying the Data Warehouse with the SQL Procedure SELECT Statement,* SUGI 23.

Lafler, Kirk Paul, and Gupta, Sunil, *Solving Business Problems with the SQL Procedure,* Data Warehouse, SUGI 21.

Lafler, Kirk Paul, *Solving Common Data Processing Problems with the SQL Procedure.*

Lafler, Kirk Paul, "Using the SQL Procedure," SUGI 18.

Lassiter, Frank, *Building Quality into the SAS System.*

Leighton, Ralph W., "Working with Arrays: Doing More With Less Code," SUGI 19.

Levine, Howard, *Using and Understanding SAS Formats.*

Levin, Lois, *SAS Programming Conventions.*

Linden, Carol, and Green, John III, "Writing Reports with SAS Software: What Are Your Options?" Observations, First Quarter, 1994.

Ma, Juliana Meimei and Schlotzhauer, Sandra, "How and When to Use WHERE," Beginning Tutorials.

Ma, Juliana Meimei and Schlotzhauer, Sandra, "Fast Track to PROC REPORT Results," Beginning Tutorials, SUGI 25.

Ma, Juliana M., *The Art of Testing Programs with an Emphasis on Larger Files.*

Mason, Phil, *In the Know: SAS Tips & Techniques From Around the Globe,* Cary, NC: SAS Institute, Inc., 1996.

Mason, Phil, "Several Ways to Tune a SORT," Bay Area SAS Users Group, May 2003.

McNeil, Sandy, "What's New in the Output Delivery System, Version 9.0," Data Presentation, SUGI 27.

Microsoft Press, "Microsoft ACCESS 2 Step by Step," Redmond, WA, 1994.

Miron, Thomas, "The Secret Life of the DATA Step," SUGI 21.

Noga, Stephen M., "The Tabulate Procedure: One Step Beyond the Final Chapter," Information Visualization.

Pass, Ray, "I'll Have the TABULATEs a la ODS Please, With a Table of Contents on the Side," SUGI 27.

Pass, Ray, *PROC Report: An Introduction to the Batch Language,* SUGI 22, Hands-On Workshop.

Peng, Joanne and Wong, Shu-Yeng, *Learning SAS by Diagrams and Examples.*

Perry, William, *Effective Methods for Software Testing,* New York, John Wiley & Sons, Inc., 1995.

Polzin, Jeffery A., *DATA Step Efficiency and Performance,* SUGI 19.

Rabb, Merry G., *Using the SQL Procedure in Data Management Applications.*

Riba, S. David, "New Tricks for an Old Tool: Using Custom Formats for Data Validation and Program Efficiency," Tutorials, SESUG 94.

Ritzow, Kim L. Kolbe, "Introduction to the UNIVARIATE Procedure," SUGI 19.

Rodriguez, Robert N., "An Introduction to ODS for Statistical Graphs in SAS 9.1," Statistical and Data Analysis, SUGI 29.

Stewart, Larry, and Fecht, Marje, "Tips and Tricks for Easier Reporting," Hands-On Workshop, SUGI 27.

Stinson, Kimberly, "Getting Up to Speed with PROC REPORT," Tutorials, SUGI 29.

Stroupe, Jane, "What's New in Versions 7 and 8 for SAS Files," *Bay Area SAS Users Group Newsletter,* June 2000.

Suhr, Diana D., Gavin, Thomas J., Kovitz, Susan B., and Goodman, Brent S., *From Madness to Method.*

Suhr, Diana, *"Hurray for Arrays!"* Proceedings of the 10th Annual Western Users of SAS Software, San Diego, CA.

Virgile, Robert, *An Array of Challenges,* Cary, NC: SAS Institute, Inc., 1990.

Walgamotte, Veronica R., "Learning the Basics: Everything Your Professor Should Have Told You," Tutorials, SESUG 94.

Whitney, C. Michael, "Taming the Chaos: A Primer on the Software Life Cycle and Programming Standards," Observations, Fourth Quarter 1996.

Winn, J. Thomas, Jr., "Advanced Features of Proc Tabulate," SUGI 24, Hands-On Workshop.

SAS Documentation

SAS Institute, Inc., *Reporting from the Field: SAS Software Experts Present Real-World Report-Writing Applications,* Cary, NC: SAS Institute, Inc., 1994.

SAS Institute, Inc., *SAS Applications Guide,* Cary, NC: SAS Institute, Inc., 1987.

SAS Institute, Inc., "SAS Certification," <http://www.sas.com/education>.

SAS Institute, Inc., *SAS Guide to Report Writing: Examples,* Cary, NC: SAS Institute, Inc., 1994.

SAS Institute, Inc., *SAS Guide to Tabulate Processing,* Cary, NC: SAS Institute, Inc., 1990.

SAS Institute, Inc., *SAS Guide to the REPORT Procedure, Reference,* Cary, NC: SAS Institute, Inc., 1995.

SAS Institute, Inc., *SAS Guide to the REPORT Procedure, Usage and Reference,* Cary, NC: SAS Institute, Inc., 1990.

SAS Institute, Inc., *SAS Language and Procedures, Usage,* Cary, NC: SAS Institute, Inc., 1989.

SAS Institute, Inc., *SAS Language, Reference,* Cary, NC: SAS Institute, Inc., 1990.

SAS Institute, Inc., *SAS Learning Edition.*

SAS Institute, Inc., SAS online documentation for Version 8.2 and Version 9.0.

SAS Institute, Inc., *SAS Procedures Guide,* Cary, NC: SAS Institute, Inc., 1990.

SAS Institute, Inc., *SAS Programming Tips: A Guide to Efficient SAS Processing,* Cary, NC: SAS Institute, Inc., 1990.

SAS Institute, Inc., *What's New in SAS Software for Release 8.2,* Cary, NC: SAS Institute, Inc., 2001.

Answers

Chapter 1.   Accessing Data—Answers to Chapter Questions

Answer 1: When you mix INFILE statement options such as FIRSTOBS = and OBS =, SAS starts reading from the record number in FIRSTOBS, 5, and stops when it reaches the record number in OBS, 10. See Example 1.10 for more information.

Answer 2: One way to correctly read missing values without a "." in the raw file is to use a COLUMN INPUT statement because it reads data from column numbers. See Example 1.2 for more information.

Answer 3: No. The flag variable created with the data set option (IN=) is temporary and is not saved with the data set. You can, however, save this variable to a new permanent variable when specifying the assignment statement. See Example 1.19 for more information about using the data set option (IN=).

Answer 4: Yes. A data set must be presorted by the same BY variables before it can be combined with another data set using the BY statement. The alternative is for the data set to be indexed by the BY variables. See Example 1.15 for more information about the SET and BY statements.

Answer 5: Yes. When using the COLUMN INPUT statement, SAS reads the data based on the order of the columns specified. See Example 1.2 for more information.

Answer 6: For common variables, the second data set has data values that overwrite data values of the first data set when merging two data sets using the MERGE statement. See Example 1.16 and Example 1.18 for more information.

Answer 7: No. When using the LIST INPUT statement, it is not possible to have embedded blanks in your data values unless the ampersand format modifier is specified. Use the COLUMN or FORMATTED INPUT statements to read embedded blanks. See Example 1.2 and Example 1.3 for more information about COLUMN and FORMATTED INPUT statements.

Answer 8: When setting two data sets together in a SET statement with one having 5 observations and another having 6 observations, the final data set will have 11 observations. SAS appends the records from the second data set to the first data set. See Example 1.14 for more information.

Answer 9: Yes, it is possible to skip rows when using the pound-n (#n) line pointer control to read raw data. SAS reads only the rows specified by the n value in the pound-n syntax. It is also possible to go back on skipped rows. The slash (/) line pointer control, however, will always read the next row line. See Example 1.7 for more information.

Answer 10: When merging two data sets with the MERGE and BY statements, the data values come from the second data set and overwrite the values from the first data set. See Example 1.18 for more information.

Answer 11: The FORMAT, INFORMAT and LABELs of common variables will be defined from the first data set in the SET statement. Note that if the FORMAT, INFORMAT or LABEL statement is not specified in the first data set, then SAS uses the FORMAT, INFORMAT and LABEL statement from the second data set if specified. When using the SET statement, the first data set in the SET statement defines the length. When concatenating data sets, make sure common variables have matching variable attributes. See Example 1.14 for more information.

Answer 12: You can use the FORMATTED INPUT statement with the mmddyy8. informat to read dates such as 10/23/78. See Example 1.3 for more information.

Answer 13: Yes. If the colon format modifier (:) is specified in the LIST INPUT statement, more than one blank or delimiter may separate the data values. See Example 1.1 for more information on LIST INPUT statements.

Answer 14: This is an INNER join. See Example 1.21 for more information.

```
/* Comparable DATA step except for common variables */
DATA TEST7;
  MERGE C (IN=INC)  D (IN=IND);
  BY PATNO;
  IF INC AND IND;
RUN;
```

Answer 15: SAS generally writes the data from the PDV to the new data set at the boundary of a DATA step. The OUTPUT statement is generally applied when using multiple INPUT statements to prevent SAS from overwriting the data in PDV with new raw data from the next INPUT statement. The OUTPUT statement forces SAS to write data values to the data set. If the OUTPUT statement is not applied, it is by default implicitly included at the end of the DATA step; thus, only the most recent data read would be written to the data set. When a single INPUT statement is used, then an implicit OUTPUT statement is already included at the end of the DATA step. See Example 1.8 for more information.

Answer 16: No. The FIRST.by_variable and LAST.by_variables are temporary variables that are not saved with the data set. You can, however, save these variables to new permanent variables when specifying the assignment statements. See Example 1.15 for more information about the FIRST.by_variable and LAST.by_variable.

Answer 17: The LEFT join combines all records from the left data set plus all matching observations from the right data set. The RIGHT join combines all records from the right data set plus all matching observations from the left data set. The FULL join combines all observations from both right and left data sets. The OUTER join is similar to the INNER join except that all records from at least one of the data sets are included in the new data set. The three types of OUTER joins are LEFT, RIGHT, and FULL. See the following examples for more information: 1.21, 1.22, 1.23, 1.24, and 1.25.

Answer 18: The equivalent DATA step for performing a FULL OUTER join is the following: MERGE with the BY statement. The results are the same except for common variables. See Example 1.25 for more information.

Answer 19: One way to read commas as data values is to use the DSD option in the INFILE statement with the LIST INPUT statement to read commas in quoted strings. See Example 1.11 for more information.

Answer 20: When using the SET or MERGE statement with the BY statement, the data set should be presorted by the BY variables. Use the SORT procedure to sort the data sets. The alternative is to have the data sets indexed by the BY variables. See Example 1.15 and Example 1.18 for more information.

Answer 21: The FORMAT, INFORMAT and LABELs of common variables will be defined from the first data set in the MERGE statement. Note that if the FORMAT, INFORMAT or LABEL statement is not specified in the first data set, then SAS uses the FORMAT, INFORMAT and LABEL statement from the second data set if specified. When using the MERGE statement, the first data set in the MERGE statement defines the length. When merging data sets, make sure common variables have matching variable attributes. See Example 1.18 for more information.

Answer 22: The SELECT clause is used in the SQL procedure to select variables. See Example 1.20 for more information.

Answer 23: Yes. When using a FORMATTED INPUT statement, variables with and without informats can be specified when combined with the LIST or COLUMN INPUT statement. See Example 1.3 for more information on the FORMATTED INPUT statement.

Answer 24: When reading data, if you do not specify the length of variables, the default length is 8 bytes for both character and numeric variables. Make sure to specify the correct length to prevent truncation of data. See Example 1.1 for more information on the LIST INPUT statement.

Answer 25: Yes, the trailing @ control can be used in the LIST, COLUMN, or FORMATTED INPUT statements. See Example 1.8 for more information.

Answer 26: d. When performing a match merge with the BY statement, SAS keeps all records with a unique combination of the BY variables. See Example 1.18 for more information.

Answer 27: By default, missing numeric data are represented with periods and missing character data are represented with blanks. You can set the system option for missing to reset the missing character.

Answer 28: You can use the SQL procedure to

1. Retrieve and manipulate SAS tables
2. Add or modify data values in a table
3. Add, modify, or drop columns in a table
4. Create tables
5. Generate reports

Answer 29: The equivalent DATA step for performing a conventional INNER join is the following: MERGE statement with data set option (IN=) for each data set, BY statement, and IF statement to include only matching BY variable values from both data sets. See Example 1.21 for more information.

Answer 30: No, the double trailing @@ control should never be used in the COLUMN INPUT statement because SAS automatically positions the column pointer as it reads data across the raw data file. See Example 1.9 for more information on the double trailing @@ control.

Answer 31: Character data are case-sensitive; thus "Sally" is not equivalent to "sally". Because of the case sensitivity, you need to specify the correct spelling or use the UPCASE() or LOWCASE() function when querying the data. See Chapter 3, section 3.5 for more information on SAS functions.

Answer 32: If the raw data file contains multiple delimiters or a single delimiter other than a comma, then you can specify the DSD option and the DLM= option in the INFILE statement to read the data. See Example 1.11 for more information on the DSD option.

Answer 33: You can expect to get similar results if using the MERGE statement and multiple SET statements to merge data sets if both the data sets have the same number of observations. This is because the multiple SET statements will restrict the new data set to contain the number of records that are in the smallest data set. If one data set has more observations than the other data set, then merging the data sets using the MERGE statement will have more records than when using the multiple SET statements. See Example 1.16 and Example 1.17 for more information.

Answer 34: The following are some reasons for using the INFILE statement option DSD with a LIST INPUT statement:

1. Allows for reading comma delimited or quoted string files
2. Sets the default delimiter to comma
3. Treats two consecutive delimiters as a missing value
4. Removes quotation marks from character values

See Example 1.11 for more information.

Answer 35: When using the SQL procedure to combine tables, the maximum number of tables that can be combined at once is 16. In Version 9.1, however, SAS allows up to 32 tables to be combined at once.

Answer 36: Only the length of character variables may be specified in the COLUMN and FORMATTED INPUT statements. Numeric variables will always by 8 bytes unless specified by a LENGTH or ATTRIB statement. See section 1.2 for more information.

Answer 37: When using the SQL procedure to combine data sets, will you get a WARNING message when selecting common variables unless you select variables from individual data sets or use the COALESCE( ) function. It is recommended to change the selection of common variables to prevent this message. See Example 1.20 and Example 1.26 for more information. SAS will not issue a WARNING message with using the MERGE with BY statement to combine data sets with common variables.

Chapter 2.   Creating Data Structures—Answers to Chapter Questions

Answer 1: To assign a date constant of January 1, 2002, specify the following statement: STARTDT = '01JAN2002'd; see Example 2.7 for more information.

Answer 2: If both DROP and KEEP statements are applied within the DATA step, then the order of priority is DROP and then KEEP. See Example 2.6 for more information.

Answer 3: No. If an SAS data set is created from another SAS data set, then SAS bypasses the input buffer and directly places the values in the PDV. See Example 2.12 for more information.

Answer 4: The following methods can be used to control the selection of observations within a data set:

1) IF–THEN statement
2) WHERE statement
3) SELECT–WHEN statement

When comparing character variables, remember to assure case insensitivity. See Example 2.4 and Example 2.5 for more information.

Answer 5: Yes. The INFORMAT statement is a compile-time statement. Compile time statements can be placed anywhere in the DATA step because they will run first and because they are independent of the execution steps, which run sequentially. See Example 2.11 and Example 2.12 for more information on the compilation phase.

Answer 6: Yes. Formats can be used in PUT statements to write to an external file. The formats are useful to specify the display of data values saved in the file. See Example 2.8 for more information.

Answer 7:

```
 proc print data =test (obs=20);
   where gender = 'Male';
 run;
```

Of the first 20 records processed by the PRINT procedure, SAS displays only the records that meet gender = 'Male' condition. See Example 2.3 for more information on data set options.

Note that if using the WHERE data set option, then make sure you are not using the same variable in the DROP data set option. For example, the following code is not allowed:

```
 proc print data=test (where= (patno=1) drop = patno);
 run;
```

Answer 8: When SAS creates data sets from external files, SAS creates and saves the variables in the order SAS reads the variable from the INPUT statement if no other compile-time statements are specified before the INPUT statement. If statements such as the LENGTH or ATTRIB statement are specified before the INPUT statement, then SAS uses these statements to define the variable attributes and order. See Example 2.11 for more information.

Answer 9: Data values are written to the output data set during the execution phase. As each record is processed in the DATA step, a record is written to the output data set. See Example 2.11 and Example 2.12 for more information.

Answer 10: The advantage of creating a permanent data set over creating a temporary data set is that the data set can be accessed anytime without having to spend resources to recreate it. Permanent data sets are referenced with a libref other than the work libref. Make sure to create a libref first. See Example 2.1 for more information.

Answer 11: The SAS data value for the date January 1, 1960 is 0. SAS stores dates as a numeric data type variable. See Example 2.7 for more information on SAS dates.

Answer 12:

1. Compile-time statements are processed first and can be specified in any order.
2. Execution-time statements are processed second and run in the order of sequence.

See Example 2.11 and Example 2.12 for more information on compile and execution phases.

Answer 13: When using the IF statement to change a variable's value conditionally, the change takes place in the PDV because the input buffer only stores the original data values. The PDV stores all original, new, and temporary variables within the new data set. See Example 2.4, Example 2.5 and Example 2.11 for more information.

Answer 14: The FILE and the PUT statements are required to write to an external file from within a DATA step. The PUT statement specifies the order and variables to write to the identified external file referenced by the FILE statement. See Example 2.8 for more information.

Answer 15: The DROP and KEEP statements are used to control the selection of variables within a data set. Use the statement that is easiest or the shortest list of variables to specify. See Example 2.6 for more information.

Answer 16: The variable list shortcut equivalent to VAR ROOT1–ROOT3; is VAR ROOT1 ROOT2 ROOT3; See Table 2.3 for more information.

Answer 17: When creating data sets, an implied OUTPUT statement exists as the last DATA step statement if not already specified. See Example 2.1 for more information.

Answer 18: If a WHERE data set option is specified on the new data set and a WHERE statement is specified within the DATA step, then SAS ignores the WHERE statement and applies the WHERE data set option. See Example 2.5 for more information on WHERE statements.

Answer 19: The descriptor portion of the output data set is created in the compilation phase. During the compile phase, all variables and their attributes are specified. See Example 2.11 for more information.

Answer 20: No. By default, compile-time statements, such as DROP, cannot be conditionally executed. All compile-time statements are executed at the same time. See Example 2.11 and Example 2.12 for more information.

Answer 21: If the compilation phase results in an error, the data set is not created.

Answer 22: To process the first record in a data set, apply the following syntax:

```
if _n_ = 1 then do; /* statement to process when reading
first observation */ end;
```

See Example 2.11 for more information.

Answer 23: No, there is not a difference between specifying the DROP= data set option before the OBS= data set option on the same data set. The order of the data set options is irrelevant. See Example 2.3 for more information on data set options.

Answer 24: No, a WHERE statement cannot include new variables or temporary variables such as _N_. This is because the subset condition is applied before the data goes into the PDV; thus, the new variable and temporary variables do not exist yet. See Example 2.5 and Figure 2.1 for more information on WHERE statements.

Answer 25: No, WHERE statements cannot be applied to DATA steps with an INPUT statement. The WHERE statement requires the SET, MERGE, or UPDATE statement if specified within a DATA step. See Example 2.5 and Figure 2.1 for more information on WHERE statements.

Answer 26: If the YEARCUTOFF= option is set to 1920, the 100-year span window SAS uses to read dates with only two-digit years is 1920 to 2019. Therefore, any dates with two-digit years between 20 and 99 will use 19 as the prefix and any dates with two-digit years between 00 and 19 will use 20 as the prefix, unless the YEARCUTOFF value is changed. The year 1920 is the default value for the YEARCUTOFF= option. See Example 2.7 for more information on SAS dates.

Chapter 3.   Managing Data—Answers to Chapter Questions

Answer 1: x=SUM(a, b, c, 0);

The SUM() and MEAN() functions use non-missing values. If all variables have missing values, then to prevent assigning a missing value to the *x* variable, include 0 in the function. See Example 3.8 for more information on SAS functions.

Answer 2: Although the ARRAY statement specifies variables *sales85* to *sales90*, the *tsales* variable calculation is based on variables *sales86* to *sales89* because of the array reference in the DO loop. The following sales variables are created unless they already exist in the y data set: *sales85, sales86, sales87, sales88, sales89,* and *sales90.* See Example 3.13 for more information on ARRAY statements.

Answer 3: The period is required in all informats and formats so that SAS can distinguish between format names and variable names. Without the period, SAS will assume the informat and format names are variable names. This can cause unexpected results, as SAS will expect these variables to exist or will create them.

Answer 4: If a SAS procedure does not have the data set name specified, then SAS uses the most recently created data set. It is always better to specify the data set name in the DATA step and all SAS procedures. This not only assures that the correct data set is processed, but also makes it easier to read the code.

Answer 5: One method to test character variable containing a numeric value is using the INDEXC() function as follows:

```
if (indexc(lname, '0123456789')) > 0
then put '*** lname contains a number ***';
```

See Examples 3.2 and 3.8 for more information on SAS functions.

Answer 6: Yes, it is possible to use the RETAIN statement to preserve a character string in a character variable without resetting to blanks for the next observation. The RETAIN statement can also preserve numeric values without resetting to missing for the next observation. See Example 3.7 for more information.

Answer 7: The SUBSTR() function is the only SAS function that can be used as a pseudo-variable to replace data values. See Section 3.5 for more information.

Answer 8: No, use the ATTRIB statement or multiple LABEL statements to assign multiple labels. See Example 3.1 for more information.

Answer 9: Arrays cannot contain both character and numeric data values. Separate ARRAY statements must be created for character and numeric variables. See Example 3.13 for more information.

Answer 10: When using an ARRAY statement without specifying the list of array elements, SAS uses the array name as the root name with a numeric suffix for each element. See Example 3.14 for more information.

Answer 11: Of the two types of DO loops, DO UNTIL and DO WHILE, the DO WHILE statement evaluates the expression at the top of the DO loop and will execute while the expression is true. Hence, if the condition is false, then the DO UNTIL will cause the loop to be executed at least once. See Examples 3.10, Example 3.11, and Example 3.12 for more information.

Answer 12: The following three methods test if the character variable *name* contains blanks:

```
1)  if name = ' ' then put '** name contains blanks **';
2)  if length(trim(name)) = 0 then put '** name contains
    blanks **';
3)  if index(name, ' ') >= 1 then put '** name contains
    blanks **';
```

Answer 13: The following code removes the first character in the *name* variable and starts on the second character value:

```
data x;
  name = substr(name, 2);
run;
```

See Example 3.8 for more information on SAS functions.

Answer 14: Multiple variables specified in FORMAT, LENGTH, and LABEL statements are separated by spaces. See Example 3.1, Example 3.2, and Example 3.3 for more information.

Answer 15: The format in the PUT() function does not always have to be a character format. The format type, however, must match the source variable type; thus, if the source variable is numeric, then the format must be numeric, and if the source variable is character, then the format must be character. The result of the PUT() function, however, is always a character variable. See Example 3.9 for more information.

Answer 16: The following character functions return a numeric value: INDEX(), LENGTH(). Most other character functions return a character value. See Example 3.8 for more information on SAS functions.

Answer 17: Without a LENGTH statement to specify the length of the new variable *y*, SAS determines the length from the first data value. Since "NO" is the first data value seen by SAS, the length of the *y* variable is 2 instead of 3 and "YE" is stored, not "YES," when *x* is equal to 2. See Example 3.3 for more information.

Answer 18: Using the _TEMPORARY_ option in an ARRAY statement creates an array of values instead of creating an array of variables. While this approach saves you from needing to delete unwanted variables, you must now reference array elements instead of variable names. See Example 3.14 for more information.

Answer 19: Since SAS sees the assignment statement before the ATTRIB statement, the length of the *x* variable is determined from the first value "toolong." The length of 7 is not reset to 4 as specified in the ATTRIB statement. In addition, SAS gives the following warning message: "Length of character variable has already been set." Use the LENGTH statement as the very first statement in the DATA step to declare the length of a character variable.

The SAS log of the DATA step in this example will display "x = toolong len = 7."

Answer 20: The three keywords available in the VALUE statement of the FORMAT procedure are LOW to define the minimum value, HIGH to define the maximum value, and OTHER. See Example 3.5 for more information.

Answer 21: Yes. The following SAS code will assign the *day* variable the values "mon," "tue," "wed," "thur," "fri," "sat," and "sun":

```
data test;
  do day = 'mon', 'tue', 'wed', 'thur', 'fri', 'sat', 'sun';
    put day = ;
  end;
run;
```

Since the final data set will contain only one record, the last assignment for the day variable, "sun," will be saved to the *test* data set. See Example 3.10 and Example 3.11 for more information on DO loops.

```
data test;
  length x 3. y $10.;
  a = x;  b = y;
run;
```

Answer 22: The LENGTH specifies the length and type of variables *x* and *y* as 3 numbers and 10 characters respectively. Since the numeric variable *a* is created without a LENGTH statement, SAS uses the default length of 8. Since the character variable *b* is assigned the value of the *y* variable, the *b* variable takes the length of the *y* variable. See Example 3.3 for more information on the LENGTH statement.

Below are the results of the CONTENTS procedure of the data set *test*:

variable	type	length
x	numeric	3.
y	character	$10.
a	numeric	8.
b	character	$10.

Answer 23: One method to test whether the character variable *lname* contains a capital letter is the following code:

```
data test;
  if indexc(lname, 'ABCDEFGHIJKLMNOPQRSTUVWXYZ') > 0
  then put '** lname  contains a capital letter **';
run;
```

See Example 3.8 for more information on SAS functions.

Answer 24: The maximum length of a variable's label is 256. By using the variable's label, you can specify the LABEL option in the PRINT procedure to display variable labels as headers instead of the variable names. See Example 3.1 for more information.

Answer 25: Format labels can be up to 200 characters long and must be enclosed in quotes. Formats are useful to control the display of variable values. See Example 3.2 and Example 3.5 for more information.

Answer 26: At the completion of the DATA step, the value of the index variable would be 11 and the value of the capital variable would be 51. See Example 3.10, Example 3.11, and Example 3.12 for more information.

Answer 27: The length of the character string that results from using most SAS character functions, such as SUBSTR(), LEFT(), and TRIM(), is the same length as their arguments. Other SAS functions such as SCAN() will return a length of 200. When working with character functions, you should make sure the result of the function is what you would expect and that it will not cause SAS to generate an error or a warning message. See Example 3.8 for more information on SAS functions.

Answer 28: The informat of the INPUT() function determines whether the type of the result is numeric or character. If the informat is numeric, then the result is a numeric variable. If the informat is character, then the result is a character variable. See Table 3.3 and Example 3.9 for more information.

Answer 29: The length attribute is specified with the LENGTH statement. This defines the amount of space allocated to save the data. The default length of all variables is 8 bytes. See Example 3.3 for more information.

Answer 30: The length attribute determines if a variable is character, such as $10, or numeric, such as 4. In addition, the informat and format attributes can also determine if the variable is character or numeric.

Answer 31: When using the RENAME statement to change a variable's name in the new data set, it is not possible to reference the new variable's name within the same DATA step. This is because the new name is only available in the new data set. The original variable's name must be used to reference the variable. See Example 3.4 for more information.

Answer 32: Although it is system dependent, the general range length for numeric variables is 3 to 8.

Answer 33: Yes, SAS functions such as SUM() and MEAN() can process multiple variables across one observation as compared to SAS procedures processing only one variable per observation. See Example 3.8 for more information on SAS functions.

Answer 34: DO loops can be used for the following:

1. To perform repetitive calculations
2. To generate data
3. To execute SAS code conditionally
4. To read data

Answer 35: The RETAIN statement is required to accumulate totals, because, by default, SAS resets variables to missing values after each DATA step iteration. The RETAIN statement preserves the current value for the next DATA step iteration so that totals can be tabulated. See Example 3.7 for more information.

Answer 36: The INDEXC() function can be used to determine if a string contains a specific letter. The INDEX() function may not determine this if there is not an exact match with the string to locate. See Table 3.2 for more information.

Answer 37: When a permanent format is applied to a variable, then by default that format is always used and displays the label value. To override this, you can apply a temporary format of x.d or $x. at the SAS procedure level to display the actual number or character data or use a different format. See Example 3.5 for more information.

Answer 38: When sorting a data set with the BY statement, the FIRST.by_variable and the LAST.by_variable variables are temporarily created in the DATA step. These temporary variables are useful to identify when the unique value of the BY variable changes. The first and last records of each unique BY variable have a value of 1 for each FIRST.by_variable and LAST.by_variable, respectively. All other records will have the value 0. See Example 1.15 in Chapter 1 for more information.

Answer 39: The purpose of the SCAN() function is to subset a text word. As opposed to the SUBSTR() function, the SCAN() function searches for tokens separated by delimiters. See Table 3.2 and Example 3.8 for more information on SAS functions.

Answer 40: No. When using the DATASETS procedure to modify variable attributes, each data set that you want to change must have its own MODIFY statement. See Example 3.18 for more information.

Answer 41: LABEL statements applied in DATA steps are permanent variable attributes while those applied in SAS procedures are temporary variable attributes. LABEL statements applied during a SAS procedure override the permanent variable labels. See Example 3.1 for more information.

Answer 42: When using the iterative DO loop, SAS creates the index variable as a permanent variable. See Example 3.10 for more information.

Answer 43: SAS works from the inside out to resolve expressions. The result of the inner function is used as the argument for the outer function. Make sure the argument is consistent with your expectations and the outer function; i.e., (apply this function second (start with this function first)). See Table 3.2 and Example 3.8 for more information on SAS functions.

Answer 44: If there is no LENGTH or ATTRIB statement or any reference to variables within a program, SAS automatically assigns a length once it reads the variable's first value in the raw data file or data set. See Example 3.1 and Example 3.3 for more information on LENGTH and ATTRIB statements.

Answer 45: The variable type for the *myword* variable is character in the *testa* data set because the first value assigned to it is a character. Because of the second assignment statement, the value of the *myword* variable is changed to 1. In addition, SAS gives the following note because of the different variable type: "NOTE: Numeric values have been converted to character values at the places given by: (Line):(Column)." You should apply the PUT( ) function to convert the number 1 to a character value before saving the *myword* variable.

The variable type for the *myword* variable is numeric in the *testb* data set because the first value assigned to it is a number. Since numeric variables can only store numbers and not character data, SAS gives the following note because it reassigns the *myword* variable to missing: "NOTE: Character values have been converted to numeric values at the places given by: (Line):(Column). 93:11 NOTE: Invalid numeric data, 'hello', at line 93 column 11. myword=. _ERROR_=1 _N_=1." If the character value is a number, then you should first convert the character to a number with the INPUT( ) function. See Table 3.3 and Example 3.9 for more information.

Answer 46: No. Within a DATA step, it is not possible to have compile-time statements be conditionally executed, because all compile-time statements must be executed. See Example 3.6 for more information on conditionally executed statements.

Answer 47: Since the iterative DO loop automatically increments the index variable, the DO UNTIL and DO WHILE statements need an additional statement to increment the index variable. See Examples 3.11 and Example 3.12 for more information.

Chapter 4. Generating Reports—Answers to Chapter Questions

Answer 1: The MISSING option in the PROC TABULATE statement includes missing values for all class variables. The PRINTMISS option in the TABLE statement includes missing values for all analysis variables. See Example 4.5 and Example 4.6 for more information.

Answer 2: The default statistic for using the analysis variable in the TABULATE procedure is SUM. The default statistic for using class variables is COUNT.

Answer 3: In the REPORT procedure, the difference between the GROUP, DISPLAY, and ACROSS options is as follows: GROUP consolidates values of several rows into one row; DISPLAY displays variables as they appear in the data set; and ACROSS uses the variables to form column headers. See Table 4.4 and Examples 4.7–4.9 for more information.

Answer 4: When using ODS to create various file types such as HTML, PDF, and RTF, SAS can simultaneously create these files without needing to rerun the SAS procedures. See Example 4.10 for more information.

Answer 5: Yes. Formats can be specified in the PRINT procedure. These formats temporarily override the permanent formats for the duration of the PRINT procedure. See Example 4.1 for more information.

Answer 6: The three dimensions when using the TABULATE procedure are column, row, and page, in that order of priority. See Table 4.2, Examples 4.5, and Example 4.6 for more information.

Answer 7: When using the REPORT procedure, if the COLUMN statement is not specified, then SAS displays the variables in the order in which they were saved in the data set. See Example 4.7 for more information.

Answer 8: By default, the MEANS procedure produces a report while the SUMMARY procedure provides an output data set unless the PRINT option is specified. Other than that, the two SAS procedures are similar in syntax and results.

Answer 9: It is possible to save the results from multiple SAS procedures in the same output file when using ODS. See Example 4.10 and Example 4.11 for more information.

Answer 10: The F= format modifier in the PROC TABULATE statement applies to all class variables while the F= format modifier in the TABLE statement can be used for analysis variables. See Example 4.6 for more information.

Answer 11: If the PRINT procedure did not have a VAR statement, then the arrangement of variables would be determined by the physical position of the variables in the data set. In addition, all variables would be listed. It is possible to see the physical position of all variables by using the CONTENTS procedure and reviewing the varnum values.

Answer 12: The default statistic for using the class variable in the TABULATE procedure is COUNT. See Example 4.5 and Example 4.6 for more information.

Answer 13: The PROC PRINT option to use variable labels in the output is LABEL. This is useful for improving the understanding of each variable. See Example 4.1 for more information.

Answer 14: In the PRINT procedure, the BY statement is used to output a separate listing for each BY group. Make sure the data set is sorted in the same order of the BY statement before the PRINT procedure. See Example 4.1 for more information.

Answer 15: It is not possible to access an HTML file created by SAS without first executing the ODS HTML CLOSE or ODS CLOSE statement. ODS must close any file it creates, including RTF, PDF, etc., before you can access it. In addition, the RUN statement must be executed before the ODS HTML CLOSE statement.

Answer 16: The data set does not need to be presorted before running the MEANS procedure if using the CLASS statement. If, however, you use the BY statement, then the data set must be sorted in advance of the MEANS procedure. See Example 4.2 and Example 4.3 for more information.

Answer 17: The default title in all output files is "The SAS System." This can be changed, however, by specifying the TITLE statement. You can have up to ten titles and footnotes in your report. See Section 4.8 for more information.

Answer 18: When using the TABULATE procedure, it is not possible to have statistics on several dimensions. All statistics must be on the same dimension. See Section 4.5 for more information.

Answer 19: The COLUMN statement in the REPORT procedure is used to specify and order all variables to be displayed. See Example 4.7, Example 4.8, and Example 4.9 for more information.

Answer 20: In general, the OUTPUT statement is used to save the results of SAS procedures to a data set. Note that not all SAS procedures may support the OUTPUT statement. An alternative to the OUTPUT statement is to use the OUTPUT destination in ODS to save the results to a data set.

Answer 21: The MAXDEC= option on the MEANS procedure controls the number of decimal places displayed. You can use this to improve the display of numeric variables since, by default, SAS displays the full width of each numeric variable. Since the MEANS procedure excludes missing values, the statistics produced are based on non-missing values.

Answer 22: The DEFINE statement in the REPORT procedure is used to define variable specifications such as LABEL and FORMAT. See Examples 4.7, Example 4.8, and Example 4.9 for more information.

Answer 23: The REPORT procedure syntax to create the table is below. See Example 4.7 for more information.

```
proc report data=drug headline headskip missing split='*'
nowindows;
  columns drug gender race dose;
  define gender / display format=$genderf. width=8 center
  'Baseline*Gender';
  define race / display width=8 center 'Baseline*Race';
```

```
  define drug / group width=7 center 'Study*Drug';
  define dose / display width=5 'Study*Drug*Dose';
 run;
```

Answer 24: When using ODS, you can use the STYLE= option to change the style of your output file. See Section 4.7 for more information.

Answer 25: The system option to prevent the display of the BY variable name and value in the title when using the BY statement with the PRINT procedure is NOBYLINE. See Section 4.2 and Section 4.8 for more information.

Answer 26: One of the benefits for specifying separate TABLE statements when using the FREQ procedure is to control the options and statistics for each table. With a single TABLE statement, any option or statistics specified will affect all variables in the TABLE statement. See Example 4.4 for more information.

Answer 27: When using the FREQ procedure to create cross-tabulation results, the order of the variables controls the layout of the results. The first variable specified defines the rows and the second variable defines the columns of the table. See Example 4.4 for more information.

Answer 28: When creating a new variable using the COMPUTE block statements in the REPORT procedure, you are required to specify COMPUTED in the DEFINE statement of the new variable. COMPUTED defines the usage of the variable. Note that new variables can be formatted, labeled, or summarized as true data set variables. See Table 4.4 for more information.

Chapter 5. Handling Errors—Answers to Chapter Questions

Answer 1: ERRORs and WARNING messages can be caused from any one of the following:

1. Incorrect syntax
2. Incorrect program logic
3. Invalid data

Answer 2: The following are some methods for debugging SAS programs:

```
1. put _all_;
2. put varname = ;
3. proc contents data=; run;
4. proc freq data=, run; proc print data=; run;
```

See Table 5.3 for more information.

Answer 3: No. The "Invalid data" note appears when SAS encounters unexpected data while reading the data using the INPUT statement. When using the SET statement, you are accessing a data set that is already created. See Example 5.16 for more information.

Answer 4: No. Only the syntax error generating ERROR messages are critical to prevent the creation of data sets. The NOTEs and WARNING syntax errors will create a data set that may contain incomplete data. See Section 5.2 for more information.

Answer 5: In general, SAS will store all the data if the length specified is large enough and will only incorrectly display the data as being truncated for character or asterisks for numeric variables if the format applied is not long enough. See Example 5.5 for more information.

Answer 6: No. SAS sets the new variable to missing only for records containing a missing value in any of the variables used in the calculation. See Example 5.12 for more information.

Answer 7: In general, SAS correctly converts numeric values from character variables when performing numeric calculations. To prevent getting the SAS NOTE and possible incorrect results, it is better to convert variable types using the INPUT() and PUT() functions. See Example 3.9 in Chapter 3 and Example 5.13 for more information.

Answer 8: When SAS cannot locate a variable specified in the DATA step within the data set, it creates the variable and sets the value to missing for each observation. It is best to assign all variables.

Answer 9: The best way to resolve the SAS NOTE "Merge statement has more than one data set with repeats of BY variable" is to specify BY variables that uniquely identify records in each data set. See Example 1.18 in Chapter 1 and Example 5.9 for more information.

Answer 10: SAS gives the ERROR message "WHERE clause operator requires compatible variables" when you are trying to use a numeric operator on a character variable or when using a character operator on a numeric variable. Make sure to match both variable and operator types in each conditional execution statement. See Example 2.5 in Chapter 2 and Example 5.4 for more information.

Chapter 6. Version 8.2 and Version 9.1 Enhancements —Answers to Chapter Questions

Answer 1: The system option that can be used to run a Version 6.12 SAS program from within SAS Version 8 or higher is VALIDVARNAME=V6 or UPCASE. Make sure your Version 6.12 SAS programs have all variable names less than or equal to eight characters in length. See Example 6.2 and Example 6.3 for more information.

Answer 2: The types of integrity constraints available are the following:

1. NOT NULL
2. VALID VALUES
3. UNIQUE

See Example 6.4 for more information.

Answer 3: The maximum number of generation data sets that SAS can create when backing up data using the GENMAX= data set option is 100. See Example 6.6 for more information.

Answer 4: The alternative method to reference the generation data set is tests#002. See Example 6.6 for more information.

Answer 5: In Version 9.1, up to 32 views or tables can be referenced in the SQL procedure. See Section 6.3 for more information.

Index

Please note that numbers with t and f stand for tables and figures, respectively.

9 781584 885016